A series of student texts in

CONTEMPORARY BIOLOGY

General Editors:
Professor E. J. W. Barrington, F.R.S.
Professor Arthur J. Willis

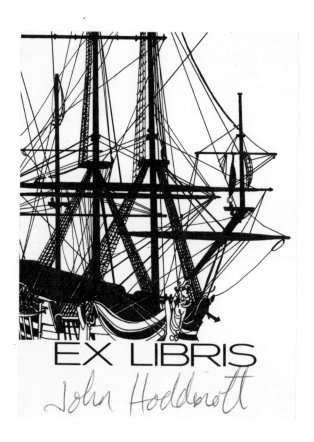

EX LIBRIS

John Hoddinott

Plant Anatomy:
Experiment and Interpretation

Part 2

ORGANS

Elizabeth G. Cutter
Ph.D., D.Sc.

Professor of Botany, University of California, Davis

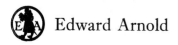 Edward Arnold

© Elizabeth G. Cutter 1971

First published 1971
by Edward Arnold (Publishers) Ltd.,
41 Maddox Street,
London, W1R oAN

Boards Edition ISBN: 0 7131 2301 x
Paper Edition   ISBN: 0 7131 2302 8

Printed in Great Britain by
William Clowes and Sons, Limited, London, Beccles and Colchester

Preface

This book forms a companion and sequel to Part 1,[127] in which the component cells and tissues of the plant body were described and discussed. In the present volume the aggregations of tissues which constitute the various organs of the plant are described and illustrated. As in Part 1, an attempt is made to relate the structure of the various plant parts to processes of growth and metabolism. Wherever it exists and is appropriate, experimental work is integrated with the more classical descriptive material of the text.

The study of plant anatomy is based on careful observation of the material. This can now be carried out at various levels, using as tools the light microscope, the transmission electron microscope and, more recently, and as yet for surface features only, the scanning electron microscope. Following the initial observations, as in other scientific disciplines, various hypotheses may be formulated. These can then often be tested by appropriate experiments. Both the initial observations, and the results of these experiments, must then be interpreted in the light of current scientific knowledge. In this volume an attempt is made to integrate the experimental and the more descriptive approaches to plant anatomy whenever possible. Because of the nature of the subject matter, illustrations have been considered to be of great importance as an aid in interpreting the material.

Although it is often believed that the anatomy of plants has been fully worked out, in fact there is a great need for a critical re-examination of various problems in this field. Interpretations of observations made decades and even centuries ago are still accepted, sometimes unquestioningly. While many of these observations are indeed correct, a fact which reflects great credit on earlier workers much less well equipped than we are today, the interpretation of them necessarily requires re-thinking in the light of advances in other scientific disciplines, or in plant anatomy itself, since that time. There is still no dearth of research problems in plant anatomy.

Gaps in our knowledge or understanding of plant structure are frequently pointed out in the present volume, and some suggestions for new investigations are made.

I am greatly indebted to many authors, including Drs. G. Bernier, N. H. Boke, C. H. Bornman, C. L. Brown, D. R. Dobbins, K. Esau, Mr. L. J. Feldman, Drs. E. M. Gifford, Jr., B. E. S. Gunning, W. Halperin, L. L. Hoefert, D. R. Kaplan, P. B. Kaufman, R. W. Kennedy, W. M. Laetsch, T. E. Mallory, W. F. Millington, P. R. Morey, J. P. Nitsch, R. L. Peterson, T. R. Pray, H. Schneider, Sister P. Schulz, Drs. B. C. Sharman, S. S. Tepfer, J. G. Torrey, Professor P. F. Wareing and Mrs. M. J. York, for supplying unpublished photographs or illustrations previously published by them, and to these and many others, and their publishers, for permission to reproduce illustrations from their publications. These sources are fully acknowledged in the captions to the figures. I am grateful to Dr. R. H. Falk and Mrs. M. J. York for their skill in taking scanning electron micrographs, to Mr. C. Y. Hung for some of the transmission electron micrographs, to Miss Sonia Cook and Mr. L. J. Feldman for making the sections from which some of the original illustrations were made, and to Mr. W. Russell for supplying some of the plant material. Dr. Bryan Truelove has again read all of the text and made many valuable suggestions. I am grateful also to my colleagues Dr. E. M. Gifford, Jr. and Dr. C. R. Stocking for their encouragement and for constructive suggestions pertaining to certain chapters, and to all of my colleagues at Davis for creating such a stimulating atmosphere in which to work. I remain responsible for all omissions and errors in the text.

My debt of gratitude to Professor C. W. Wardlaw, who has nurtured and stimulated my interest in developmental problems over a number of years by his teaching and writings, and by many friendly discussions, is incalculable. Certainly my approach to the study of plant anatomy owes a great deal to his influence. I greatly appreciate his ready advice and constructive criticism. Sincere thanks are also due to my graduate students for many pleasant discussions and useful suggestions, and to other graduate and undergraduate students at Davis who have participated in the course in plant anatomy and have helped to shape it, and this book, by their questions and comments.

Lastly I am indebted once again to Professor A. J. Willis, general editor of this series, for his wise and constructive criticism and both to him and to my publishers for their assistance at all times and for their enduring patience.

Davis,   E.G.C.
1970

Acknowledgements

I am indebted to the following publishers, scientific journals and other copyright holders for permission to reproduce the figures indicated:

Academic Press, Inc. (Figs. 4.7, 8.14); Académie Royale de Belgique (Fig. 6.12); *Açta Botanica Neerlandica* (Fig. 7.9); *Agricultural Publications, University of California* (Figs. 7.10, 7.11); American Association for the Advancement of Science (Figs. 6.7, 8.15); *American Journal of Botany* (Figs. 2.7, 2.9, 2.22, 3.7, 3.9, 4.9, 4.11, 4.12, 5.9, 5.10, 5.11, 5.12, 5.13, 5.14, 5.15, 5.16, 5.17, 5.18, 5.20, 5.31, 5.32, 5.39, 5.40, 5.48, 5.56, 6.13, 6.14, 6.18, 6.19, 6.20, 6.24, 6.28, 7.2, 7.3, 8.5, 8.7, 8.8, 8.9, 8.13); G. Bell and Sons, Ltd. (Figs. 2.1, 2.12); A. and C. Black (Figs. 4.4, 4.6); Blackwell Scientific Publications (Fig. 5.38); *Botanical Review* (Figs. 3.6, 6.29); Brookhaven National Laboratory (Figs. 6.7, 6.11); *Bulletin of the Torrey Botanical Club* (Figs. 5.49, 5.50, 5.51); Butterworths Scientific Publications (Fig. 5.52); Clarendon Press (Figs. 2.17, 3.5, 3.12, 4.10, 5.53, 7.12); Commonwealth Scientific and Industrial Research Organization, Australia (Figs. 4.10, 6.15); *Endeavour* (Figs. 2.3, 3.1); Dr. L. T. Evans (Fig. 6.15); Gustav Fischer (Fig. 3.2); W. H. Freeman and Company (Fig. 6.22); Mrs. Grace Grattan (Fig. 3.2); Linnean Society of London (Figs. 3.6, 5.47, 7.16, 7.17); McGraw-Hill Book Company (Figs. 2.1, 2.2); National Research Council of Canada (Figs. 2.15, 2.16); *New Phytologist* (Figs. 2.4, 2.5, 3.2, 4.10); *Phytomorphology* (Figs. 2.22, 5.27, 7.13, 7.14, 7.15, 8.12, 8.15); *Plant Physiology* (Figs. 8.2, 8.3); Ronald Press (Fig. 7.1); Royal Society (Fig. 3.2); Society for Experimental Biology (Fig. 2.21); Springer-Verlag (Figs. 3.2, 4.14, 4.15, 4.16, 5.42, 8.6, 8.10); Syracuse University Press (Fig. 4.13); University of California Press (Figs. 6.16, 6.17, 6.32); University of Chicago Press (Figs. 5.21, 5.54, 5.55, 6.10, 7.4, 7.6, 7.7, 7.8).

Table of Contents

I

Introduction

The plant body is made up of a number of *organs*: root, stem, leaf and flower. The last itself comprises several different kinds of lateral organ (sepals, petals, stamens, carpels and sometimes also sterile members). Each organ, in turn, is made up of a number of *tissues*, each of which comprises many *cells* of one kind. In Part 1 of this work,[127] the structure of cells and tissues was discussed; in this volume attention is focussed on the organs. Much the same tissues may be present in each organ, but their arrangement with respect to one another is different. In each organ there are interactions between the various tissues. Each usually plays a functional role in the organ of which it is a part, and to a degree the principle of division of labour is exemplified by the various component tissues. The development of the component parts of the plant body from the embryo was summarized in Chapter 1 of Part 1.

This book is concerned with the organization, or orderly arrangement, of tissues into the plant organs, and especially with how this is brought about in each case. The approach to plant anatomy outlined in this book lays stress on the complex relationships between the growth of a plant, or an organ, and its structure. To this end, use is made wherever possible of the results of experiments designed to investigate the factors controlling plant structure. However, as yet, this approach is necessarily somewhat limited, because of the paucity of experimental work in some fields. That which exists has largely been done by developmental physiologists, rather than anatomists; it is hoped that this book and the previous one[127] may prompt more anatomists to pursue these lines of investigation. It has been pointed out elsewhere[513] that plant anatomy has the reputation of being a rather static, 'worked out' subject and that if the discipline is to survive as a vital science its basic concepts must be constantly re-examined and challenged. Thus the student should constantly question what he reads and try to devise new methods of investigating familiar problems.

No attempt has been made to survey the complete literature dealing with the structure of plant organs. An effort has, however, been made to cover much of the literature dealing with relevant experiments, largely because this has not been assembled together elsewhere. If it appears that recent work has been stressed to the detriment of the work of older, classical authors, this does not stem from a lack of respect and admiration for their work, but rather arises from a belief that these references can more readily be obtained elsewhere. Students should realize, however, that the descriptive anatomy carried out by Nehemiah Grew and the other classical anatomists, and by more recent workers in the field, forms an essential basis for all subsequent work. Their achievements with often primitive equipment are greatly to be admired and even emulated. The proper understanding of the basic structure of a plant or organ is essential for the interpretation of any subsequent experimental work upon it.

Interactions between the structure and metabolism of a plant are many and reciprocal. For example, vascular tissues are involved in, and presumably necessary for, the transport of nutrients and hormones within the plant; on the other hand, these very substances are capable of inducing the formation of vascular tissues in undifferentiated tissue.

In the course of the following chapters the reader will be struck by the multiplicity of the effects of **auxin** on plant structure. That these various effects are possible must thus be related to the diverse systems on which the auxin is acting, and to other factors which may also be operative. A cogent analogy was once drawn between the effects of auxin on plant tissues, and the effects of placing a coin in a slot machine. In such a case, the same coin can elicit a packet of cigarettes, a bar of chocolate, some therms of gas for heating, some electricity, packets of chewing gum, a soft drink, a carton of milk, and so on, according to the type of machine into which the coin is inserted and which is activated by it. Thus various systems can also be activated by auxin, and the end result will depend not only on the hormone but also on the system—perhaps especially the genome—on which it acts. At present, the results of many experiments are difficult to interpret because of our ignorance of the experimental systems themselves, and especially of the endogenous growth substances present. Much further effort will be required, by both anatomists and physiologists, before the various factors affecting plant structure are understood. Nevertheless, the experimental approach to plant anatomy has already been a fruitful one and has considerable future potentialities.

TERMINOLOGY

Certain terms are used frequently in the text, and are therefore defined here (see Fig. 1.1).

Proximal   —situated near or towards the point of attachment of an organ.

Distal   —situated away from the point of attachment.

Basipetal   —from the apex towards the base; for example, differentiation may occur basipetally.

Acropetal   —from the base towards the apex.

Anticlinal   —used to describe a cell wall formed at right angles to the surface of the organ. This is an anticlinal wall.

Periclinal   —used to describe a cell wall formed parallel to the surface of the organ.

Plastochrone —the time between the formation of one leaf primordium and the next.

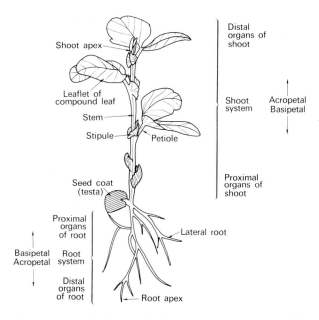

Fig. 1.1 Young plant of *Vicia faba* showing the various organs and illustrating the meaning of certain terms. $\times \frac{1}{3}$. (From Cutter, Part 1,[127] Fig. 1.3, p. 5.)

The following abbreviations are frequently used throughout:

DNA —deoxyribonucleic acid

RNA   —ribonucleic acid
IAA   —indoleacetic acid
GA   —gibberellic acid
TIBA   —2,3,5-tri-iodobenzoic acid

For the dimensions of cells it is convenient to use the micron (μ), $1/1000$ mm, as a unit of measurement. With increasing acceptance of SI units (Système International d'Unités), the micron is now written μm, the micrometre (10^{-6} m), and this practice is adopted here.

2

The Root

Almost all vascular plants have roots and, although varying considerably in form, they fulfil the common functions of absorption of water and solutes, and anchorage of the plant to the substratum. They may also act as organs for the storage of food materials, and as perennating structures.

Many dicotyledonous seedlings have one main root, the **tap** root (Fig. 2.1a), which develops from the radicle of the embryo, and may later bear lateral roots, but in some dicotyledons and many monocotyledons the tap root is soon replaced by a **fibrous** root system comprising many **adventitious** roots (Fig. 2.1b). Adventitious roots are usually formed from the tissues of stems or leaves, i.e. they occur in any position other than that of a tap root or lateral root. Recently 'long' and 'short' roots of the same species, having different potentialities for growth and development, have been reported.[420,425] More specialized types of roots also occur: for example, the **aerial** roots of many epiphytes; the **prop** roots of mangroves and other plants, which act as supporting organs; and the **contractile** roots typical of many bulbous species, among others, which by their contraction pull the shoot deeper into the soil. Some mangroves also have 'breathing roots', branch roots which grow vertically upwards from the swampy ground in which these plants thrive, thus reaching the air above.

Roots sometimes have a symbiotic relationship with fungi. This association of the fungus and the root, called **mycorrhiza**, is considered to be beneficial to both partners. This topic is dealt with more fully in another book in this series.[495] The fungal hyphae may form a weft external to the root (ectotrophic mycorrhiza), or may live within the cells of the root (endotrophic mycorrhiza). Certain nitrogen-fixing organisms also live in

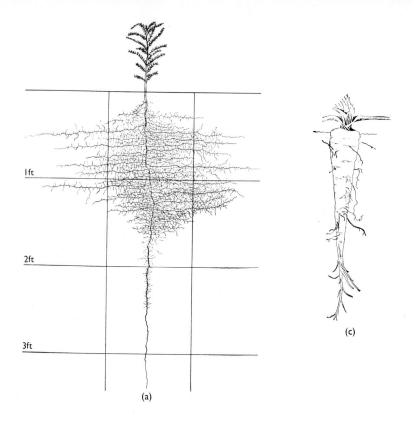

(a)

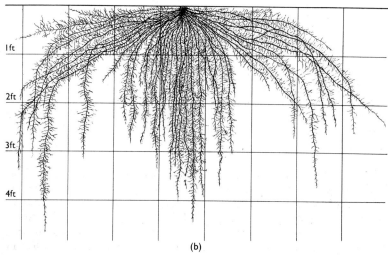

(b)

(c)

symbiosis with roots; for example, species of bacteria are responsible for the formation of the root nodules characteristic of the Leguminosae and some genera from other families.

Roots specialized for *storage*, such as carrot, *Daucus* (Fig. 2.1c), and radish, *Raphanus*, often show somewhat anomalous structure; they may comprise not only the radicle but part of the hypocotyl of the young plant. The tuberous roots of some tropical plants are important foodstuffs in certain countries, e.g. cassava (*Manihot*), yam (*Dioscorea*), and sweet potato (*Ipomoea*). Tapioca is obtained from roots by a particular treatment.

The great majority of roots grow below ground, and lack chlorophyll. The root system may be much more extensively developed than the aerial parts of the same plant. Most roots bear unicellular root hairs, which are often ephemeral. Environmental conditions in the soil may affect the form of the whole root system. In general, rather shallow root systems develop in wet, poorly aerated soils, and deeper, more extensively branched ones in drier soils. The shallow, much-branched root systems of certain cacti enable them to benefit from slight showers falling on the dry ground. Some desert plants have large tap roots capable of storing considerable quantities of water; one such root weighed 159 lb.[146]

This sometimes considerable growth and differentiation of the root is controlled by the activity of the *apical meristem*, a region of meristematic tissue which is not terminal, but is covered by the tissue of the protective *root cap*. The principal region of elongation in the root lies a short distance behind the tip; just proximal to the region of elongation is the zone of *root hairs*. *Lateral roots* are usually formed at some distance from the apical meristem. In certain species, buds are formed adventitiously from roots; where these originate from the same tissues as the primordia of lateral roots[52, 149] good material is available for the study of factors controlling the organization of root and shoot meristems.

ORGANIZATION OF THE ROOT APEX

During the later stages of development of the embryo, the cells at the root pole become arranged in a pattern characteristic of the species. This group of cells comprises the *apical meristem* of the primary root; the

---

Fig. 2.1 Types of root system. (a) Seedling of *Gleditsia triacanthos* about 3 months old, showing a tap root with numerous laterals. (b) Root system of *Zea mays* at 8 weeks old, showing numerous fibrous adventitious roots. (c) Storage root of carrot. ((a) and (b) from Weaver, J. E. (1926), *Root Development of Field Crops*, McGraw-Hill, New York. Figs. 16 and 84, pp. 49 and 182. Copyright 1926, J. E. Weaver: Used with permission of McGraw-Hill Book Company; (c) from Fritsch, F. E. and Salisbury, E. J. (1961), *Plant Form and Function*, Bell and Sons, London, Fig. 65, p. 113.)

cells are all relatively undifferentiated and meristematic, that is to say, densely protoplasmic and with large nuclei, and, initially at least, they all undergo active division. The cells of the root apical meristem are convenient objects of study with the electron microscope, and much of our early knowledge of the fine structure of plant cells was derived from this tissue region. Such studies with the electron microscope reveal that even cells of the apical meristem have some vacuoles, but these are usually not large. By various processes of growth and differentiation, described below, the tissues of the mature root are eventually derived from a number of these cells of the apical meristem, which are termed *initials*. The factors which control this organized growth are not yet understood.

In general, the apex of the root is less complicated than that of the shoot because (in the vast majority of species, at least) its lateral organs are not formed in the apical meristem but in the differentiating tissues at some distance behind it. Another major distinction between root and shoot apices is the position of the apical meristem; in the shoot this is terminal, although it may be over-arched by the leaf primordia, whereas in the root it is sub-terminal, being actually covered by the more vacuolated tissues of the root cap. Because of the absence of lateral primordia, there are apparently no regular, rhythmic changes in size and form of the root apex such as occur in most shoots (see Chapter 3). Growth is thus considered to be relatively uniform, and the developing root has been widely used in recent studies of cell growth and differentiation and in studies of enzyme distribution and activity. The requirements of roots for nutrients and growth factors have been extensively investigated by the technique of excised root culture.[72] These studies have also revealed the phenomenon of ageing of the apical meristem in excised roots; there is some evidence that in the whole plant, grown in the light, this ageing is combated by an effect of the shoot.[584]

Theories of apical organization

By a careful study of the distribution of cells in longitudinal sections of root apices, it is possible to attribute the derivation of certain files of cells to single cells or to groups of initial cells in the apical meristem. In the roots of the vascular cryptogams, e.g. the broad shield fern (*Dryopteris*), a single tetrahedral *apical cell* is present (Fig. 2.2a); it is generally considered that by its division this gives rise to all the tissues of the root. This view was the basis for the *apical cell theory* of Nägeli. Two theories have been advanced to explain the cellular patterns in the roots of flowering plants, which do not have single apical cells. The *histogen theory*, which was put forward by Hanstein[240] in 1868, postulates the existence of three cell-initiating regions, or histogens, in the apical meristem. These are the *dermatogen*, *periblem* and *plerome*, which respectively give rise to the epidermis, cortex and vascular cylinder of the mature root. This theory, which was formerly also applied to shoot apices, thus attributes a specific destiny to the derivatives of the three regions. Many studies of root meristems have described them in terms of the layers of initial cells or histogens, and roots have been divided into a number of types on this basis[273] (see also Popham[410]). This kind of classification is of descriptive value, although it is certainly inadvisable to consider such variable structures as apical meristems as conforming too rigidly to a limited number of types. In addition to that with a single apical cell, three such types are illustrated diagrammatically in Fig. 2.2. In b, one layer of cells gives rise to the

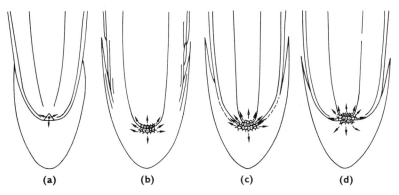

(a)   (b)   (c)   (d)

Fig. 2.2 Types of root apex. (a) Single apical cell; the cap is distinct, but has a common origin with the other tissues. (b) Two groups of initials, giving rise to the vascular cylinder and to the cortex, epidermis and root cap. (c) Three groups of initials, giving rise to vascular cylinder, cortex, and epidermis and root cap respectively. (d) Three groups of initials, giving rise to vascular cylinder, cortex and epidermis, and root cap. (From Eames, A. J. and MacDaniels, L. H. (1947). *An Introduction to Plant Anatomy*, McGraw-Hill, New York, 2nd edition, Fig. 40, p. 75. Copyright 1947 Eames and MacDaniels. Used with permission of McGraw-Hill Book Company.)

vascular cylinder, another to the cortex, epidermis and root cap, as in many gymnosperms. In c, perhaps the commonest type in dicotyledons, the cortex is formed from a separate layer of initial cells but the root cap initials are not distinct from those of the epidermis, and in d, common in monocotyledons, such as maize (*Zea mays*), the root cap has an independent origin. In such roots the cap is formed by a fourth histogen, the **calyptrogen**. In Guttenberg's terminology,[228] types c and d exhibit the 'closed' type of organization; in other species, all the tissues, or all but the vascular cylinder, arise from a common group of cells, exhibiting what is described as the 'open' type of organization.

The second theory is the ***Körper-Kappe theory***, proposed by Schüepp[454] in 1917. Since the root changes in diameter during growth, there are various points at which a single longitudinal file of cells has become a double file as a result of cell division. At these points a cell must first have divided transversely and one of its daughter cells must then have divided longitudinally. This sequence was called a T division, since the cell walls form a configuration resembling the letter T. In some parts or zones of the root, mainly in the centre, the bar or capital of the T faces the root apex, in others it faces away from the apex (Fig. 2.3). These zones of

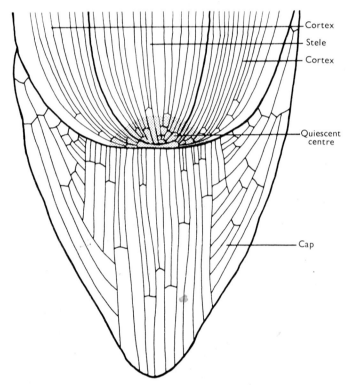

Fig. 2.3 Pattern of cell lineages in the root apex of *Zea mays*. The cortex, vascular cylinder or stele, and root cap are indicated. The sequence of divisions, with the bar of the T facing towards or away from the root apex (see text), is shown. (From Clowes,[102] Fig. 5, p. 9.)

the root, delimited by the planes of cell division, were called ***Körper*** (body) and ***Kappe*** (cap) respectively, and roots were classified according to the position of the boundary between them. This theory is thus comparable with the tunica-corpus theory in the shoot apex (see Chapter 3), being based solely on planes of cell division, and, as in the shoot, the boundary between the two zones is not constant but may vary even in roots of the same species.

By means of these theories, root apices can be adequately described in terms of the planes of cell division; this conveys some information about growth that has already taken place, but tells us little or nothing about dynamic events in the apical meristem currently occurring during growth, or about rates of cell division in the various regions of the apex. This information can often best be obtained by experimental methods. Actually, rates of nuclear division (mitosis) are usually measured; cell division (cytokinesis) generally ensues. Recent studies of the root apex have profitably combined the orthodox methods of classical plant anatomy with various modern physiological and biochemical techniques.

The promeristem

The **promeristem** can be defined as that part of the root apex which is capable of giving rise to all the tissues of the root.[96] In the roots of vascular cryptogams, for example, the promeristem would consist of the apical cell only, and in flowering plants it would comprise the initials of the histogens. There is thus a tendency to regard the promeristem as a rather small region, situated terminally in the root apical meristem, below the root cap. A growing body of modern work, however, suggests that in many roots the promeristem is broad and consists of a somewhat cup-shaped group of cells

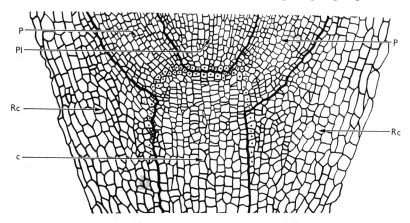

Fig. 2.4. Diagram constructed from L.S. apex of a tap root of beech, *Fagus sylvatica*. The approximate boundaries between plerome (Pl), periblem (P), root cap (Rc) and columella (c), the central part of the root cap, are shown by heavy lines. The initial cells of the promeristem are indicated by dots. (After Clowes,[85] Fig. 7, p. 261.)

on the periphery of a central inactive region. This grouping of the initial cells of the promeristem was suggested by Clowes[85] on the basis of an orthodox anatomical study of the root apex of beech, *Fagus sylvatica* (Fig. 2.4). Other workers, however, believe either that there is a small group of initials, comprising three cells or three tiers of three, or that there is a single central cell from which the initials of the histogens are in turn derived.[69, 227]

In order to investigate the reality of his suggested broad promeristem in *Fagus sylvatica*, Clowes[86] carried out various surgical experiments on seedling tap roots of this species and of broad bean, *Vicia faba*. Incisions of various kinds, and varying in depth, were made through parts of the promeristem (Fig. 2.5). Some of the cuts penetrated the pole (terminal region) of the plerome or future vascular cylinder, others did not, and some excised it completely. The cut surfaces were smeared with lanolin, and the roots were maintained in moist *Sphagnum* moss until they had grown 100–200 mm, when they were fixed and sectioned in order to study the regeneration of the tissues.

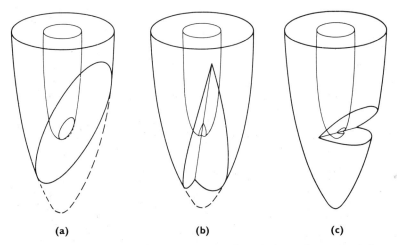

(a)   **(b)**   **(c)**

Fig. 2.5 Diagrams to show the three types of surgical experiments carried out on roots of *Fagus* and *Vicia*. (**a**) Excision of a segment of the root apex by a single oblique cut; (**b**) The removal of a vertical sector; (**c**) Removal of a horizontal wedge-shaped piece. In (**a**) and (**b**) part of the terminal region of the plerome of the future vascular cylinder is cut away. (From Clowes,[86] Fig. 1, p. 49.)

Where the plerome pole had been completely removed, no regeneration occurred. Both roots in which the cut passed into the plerome pole, and those in which the cut left this intact, regenerated, but the amount of abnormally organized tissue was greater after the former treatment. Both of these treatments, however, must have damaged some, but not all, of the initials, and since it would be almost impossible to damage only part of a promeristem comprising a small group of initials or a single central cell, Clowes considered that the results of these experiments supported the concept of a broad promeristem. Comparable experiments on root apices of grasses gave similar results, suggesting that in these species also the promeristem must be broad.[87] Other workers who have bisected root apical meristems longitudinally have also concluded that the number of initial cells must be large.[28, 398] The number of cells in the promeristem of *Vicia* has been estimated at between 40 and 50.

The quiescent centre

Especially in roots with a very regular arrangement of cells in the apical meristem, such as those of maize, *Zea mays*, it is possible to deduce from a study of the cell lineages that there is a central region of cells which divide rarely or not at all. The cells on the periphery of this hemispherical or cup-shaped region are meristematic and may be regarded as the constituents of the promeristem.[85]

By various techniques, the existence of a central inactive region, or **quiescent centre**, has now been convincingly demonstrated in the root apices of a considerable number of species (Fig. 2.3). *A priori*, this concept is at variance with many of the earlier views, which supposed that the initial cells occupied a terminal position in the apical meristem (see Fig. 2.2). However, embryonic primary roots and young lateral root primordia do not have a quiescent centre; in early growth all their cells are meristematic. The quiescent centre develops during the ontogeny of the root.[91] A quiescent centre is considered to be absent from roots with a single apical cell; however, recent work indicates that the apical cell itself may eventually become polyploid and inactive.

In 1956 Clowes[88, 89] used various techniques to study the distribution of nucleic acids in root apical meristems. By differentially staining for DNA and RNA, he was able to show that there was a central region in the roots of *Zea* where the cytoplasm had a lower content of RNA and where the cells had smaller nucleoli. By supplying roots with solutions of phosphate labelled with the radioactive isotope ^{32}P, or with adenine containing ^{14}C, Clowes was also able to demonstrate that the cells in the quiescent centre did not actively synthesize DNA. Phosphate and adenine are incorporated into both DNA and RNA, so in order to study the distribution of labelled DNA, the RNA had first to be removed from the sections by hydrolysis. In subsequent experiments[93, 97, 99] thymidine labelled with a radioactive isotope of hydrogen, tritium (^{3}H), was used; this is incorporated specifically into DNA. After these treatments, the roots were fixed, sectioned and placed in contact with autoradiographic stripping film or emulsion, which was later developed and fixed. This technique depends on the fact that certain types of photographic emulsions (e.g. X-ray film) are sensitive to radiations emitted by radioactive substances, just as they are sensitive to light. Since the film lies directly over the sections the image, or auto-radiograph, produced in the emulsion by the radiations is superimposed on the cells beneath, and the precise sites of radioactivity can be identified. In such autoradiographs silver grains, which appear black in the developed film, are grouped over labelled regions, in this instance over nuclei which were actively synthesizing DNA. In roots treated in this way a central region of the meristem where there are no labelled nuclei can be clearly distinguished (Fig. 2.6). The cells of this quiescent centre therefore

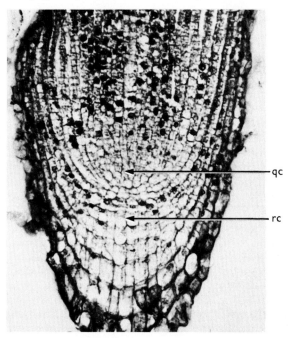

Fig. 2.6 Histoautoradiograph of L.S. root tip of radish, *Raphanus sativus*. The root was treated for 8 hours in a solution of ³H-thymidine at 1·3 μc/ml. Nuclei which have synthesized DNA during the period of treatment show as black regions (silver grains in the emulsion). The quiescent centre (qc) is evident as a region of cells with unlabelled nuclei. The nuclei of the adjacent initials of the root cap (rc) are labelled. × 200.

apparently do not synthesize DNA during the period of treatment and rarely or never undergo mitosis; in the normal development they probably play little or no part in the construction of the root.[93] Other workers have also demonstrated a quiescent centre, by the use of tritiated thymidine in a similar manner,[361] by counting the number of grains in the emulsion over nuclei after treatment with tritiated thymidine, thus obtaining a semi-quantitative result,[420] and by studying the distribution of mitotic figures in the meristem.[235]

The physiological and cytological properties of the cells in the quiescent centre have now been studied in a number of species. The cells in this region have a lower concentration of DNA, RNA and protein than any other cells in the root apex.[92, 277] They also have fewer mitochondria, little endoplasmic reticulum, and the smallest dictyosomes, nuclei and nucleoli.[105] They are less sensitive to radiation damage than other cells of the meristem,[94, 100] and more recent work has shown that the reason for this probably is that the cells of the quiescent centre are maintained for long periods in the pre-synthesis (G_1) phase of the mitotic cycle.[101–103] In roots of maize induced to become dormant by cold treatment and then restored to higher temperatures the cells of the quiescent centre were uninjured and were also stimulated to divide. It was again concluded that they escaped injury because they were in the G_1 phase of the mitotic cycle.[106]

Both the surgical experiments discussed above and experiments in which roots were subjected to various types of irradiation[94, 98] or cold treatment[106] have shown that the cells of the quiescent centre can divide under certain circumstances. Clowes[93, 96] therefore believes that the inactivity of these cells is due to their position within the apical meristem, and not to any inherent inability to undergo mitosis. The function of the quiescent centre may be to provide a reserve block of diploid cells within the root which, by virtue of their quiescence and the fact that they are maintained for long periods in the pre-synthetic phase of mitosis, are less subject to injury from agents affecting dividing cells.[94, 101, 102] It is also possible that the quiescent centre may be the site of hormone synthesis (see below), in which case its importance in the growth and development of the root would be considerable, though it would still play little or no direct role in its cellular construction.

Some recent discussion has centred on the question of whether or not this quiescent region in the meristem is to be regarded as a permanent and more or less static group of cells. For example, Guttenberg[229] argues that since a quiescent centre is often not present in young roots, and in older roots the cells of the cup or promeristem may be restored from the centre, the central cells do not constitute a 'quiescent' but rather an 'intermittent' centre, i.e. one which is only intermittently quiescent, except in old roots where no divisions can be found centrally. In *Euphorbia esula*, by means of a variation of the tritiated thymidine technique, some evidence was gained for the existence of a quiescent centre in the meristems of long roots, but not in those of the short roots of the dimorphic root system. From this finding it is argued that quiescence is superimposed upon the organization of the root apex, that a quiescent centre may or may not be present, and that if present it may vary in size.[420] In soybean, also, the quiescent centre was found to vary in size, increasing during early development of the root and then decreasing rather soon, suggesting that in this species the cells of the quiescent centre constituted a constantly changing population.[361] It has long been considered[93] that the quiescent centre is a consequence of the geometry of the root apex, and these observations seem merely to implicate it more closely in the dynamic processes of a rapidly developing organ. There is good evidence that a quiescent centre exists in the apical meristems of fully developed roots of many species; further studies may profitably be directed towards elucidating its function.

Rates of mitosis

As might be expected on the basis of these observations, cells in the different regions of the root apex divide at different rates. In maize, the initials of the root cap divide every 12 hours, whereas the value for the cells in the quiescent centre is about 200 hours. Cells in other regions have inter-mediate values.[97] In white mustard, *Sinapis alba*, the average rate of division of cells in the initials of the root cap is about once in 35 hours, compared with 500 hours in the quiescent centre.[99] This disparity in rates of division between the cells of the quiescent centre and other regions of the

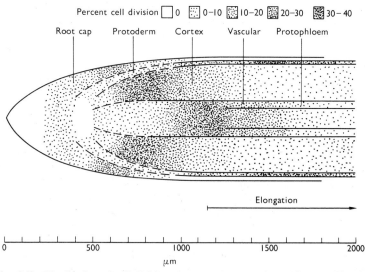

Percent cell division ☐ 0   ▦ 0–10   ▦ 10–20   ▦ 20–30   ▦ 30–40

Root cap   Protoderm   Cortex   Vascular   Protophloem

Elongation

0   500   1000   1500   2000

μm

Fig. 2.7 Distribution of cell divisions in the onion root tip at 12 noon. The data were obtained from several roots. The value 0 per cent applies only to roots counted at 12 noon and does not imply that these cells may never divide. (From Jensen and Kavaljian,[284] Fig. 4, p. 370.)

root is attributed to the much greater length of the G_1 (pre-synthesis) phase of the mitotic cycle in the quiescent centre.[104, 508] The various phases of the mitotic cycle are more fully discussed in the book by Street and Öpik[497] in this series. It is interesting to note that in roots treated with hydroxyurea, a substance which causes cells to cease dividing, treatment with indoleacetic acid (IAA) stimulated the cells of the quiescent centre to divide faster than normal by shortening the G_1 phase. This effect is a preferential one on the quiescent centre, since the rate of division in the other cells of the meristem was decreased by the treatment.[32]

In many roots maximum mitotic activity, taken over the whole width of the root, occurs some distance behind the promeristem. In onion root tips the maximal rate of division in all the tissues is at a distance of about 1 mm from the tip; the distance varies in the individual tissues of the root (Fig. 2.7).[284] In *Phleum* roots, the maximum rate of division in the epidermis occurred some 150–200 μm behind the apex.[217] Frequent collections of root tips of *Melilotus* revealed a very clear diurnal periodicity in mitosis; mitoses were much more frequent at noon and midnight than at any other time. The diurnal rhythm was found in each tissue of the root.[4] In general, at increasing distances from the root tip cell enlargement predominates, and eventually this is the only, or at least the principal, method of growth.

Root tips are usually used to demonstrate the stages of mitosis, since many of the cells are undergoing active and relatively rapid division. It is interesting that activity resembling that of the cell division factor kinetin has been found in extracts of root tips;[577] there is also good evidence that kinetin-like substances move from the root system into the shoot.[308] The structural effects of physiological interactions between the root and the shoot will be discussed further below.

TISSUE DIFFERENTIATION

The initial cells of the promeristem, which, at least in older roots, lie around the periphery of the quiescent centre, divide in such a way that the inner daughter cells remain meristematic and the outer daughter cells, after further divisions, differentiate to give rise to the various tissues of the root. The cells of the apical meristem are typically diploid, but in more mature tissues many of the cells are polyploid. The factors which control the differentiation of these genetically similar meristematic cells into the various tissues are only beginning to be understood, but it is possible in some instances to correlate the differentiation of particular tissues with specific physiological situations in the cells, although a causal connection has not always been demonstrated.

The **root cap** occupies a terminal position in the root; its cells become progressively more differentiated as distance from the root tip increases.

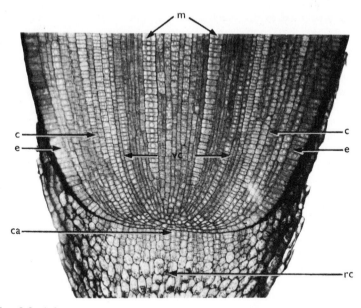

Fig. 2.8 L.S. root apex of *Zea mays*, showing differentiating tissue regions. c, cortex; ca, calyptrogen; e, epidermis; m, differentiating metaxylem; rc, root cap; vc, vascular cylinder, the boundaries of which are indicated by the arrows. × 125.

The three principal tissue systems of the root, the **epidermis, cortex** and **vascular cylinder**, become visually delimited close behind the root apex (Fig. 2.8), partly as a consequence of the onset of vacuolation in the cells of the developing cortex; the cells also differ somewhat in shape. The vascular cylinder is then evident as a central region of densely staining, small-celled tissue.

Root cap

The root cap consists of parenchymatous cells in various stages of differentiation. Because of its terminal position, the root cap has usually been considered to be a protective structure. Recent experiments have indicated, or confirmed, however, that the root cap has another function that is physiologically of great importance. In maize and barley, it is possible to remove the root cap, leaving the rest of the root intact. This has no effect on the growth of the root, but the root no longer reacts to the stimulus of gravity. After removal of the root cap, the cells of the quiescent centre divide rapidly and a new root cap is regenerated. The root cap is apparently the site of the perception of gravity; thus it appears to be capable of controlling the production in the meristem of the growth-regulating substances involved in **geotropism**, or their movement, although since root elongation continues in its absence the root cap itself is probably not the site of synthesis of growth substances.[298] In other experiments it was shown that if pea root tips were decapitated to a distance of 0·5 mm, which included the whole root cap, they failed to respond to gravity. Again, elongation of the root was not affected. Application of tri-iodobenzoic acid (TIBA), which is believed to inhibit transport of auxin, to intact root tips prevented geotropic curvature for about 3 hours, and decreased it for a longer period.[318] It is considered that this treatment probably prevented a lateral redistribution of an endogenous growth substance in response to gravity. That lateral distribution of applied auxin occurs in horizontally maintained root tips was demonstrated in experiments in which IAA labelled with ^{14}C was applied; subsequent counts of radioactivity in longitudinally halved roots showed a greater activity in the lower halves. Decapitation of 0·5 mm or more of the root tip almost entirely prevented this transverse distribution of auxin.[317] Thus these experiments are all compatible with the view that the root cap can control the movement, if not also the synthesis, of endogenous auxin in the root apex. Clearly this would be a very important function. Experiments in which only half of the root cap was removed suggested that it was the source of a growth inhibitor.[202a]

Electron microscopic studies of the distribution of organelles in root cap cells of roots subjected to geotropic stimulation indicated that only the **amyloplasts** (starch-forming plastids) showed any appreciable displacement to the lower side of the cells. It is inferred from this that the amyloplasts directly induce the transverse polarity of the cells, probably by acting on membranes near the surface of the protoplast.[221] In this connection observations on the corn mutant known as *amylomaize* are of interest. This mutant has much smaller amyloplasts than the wild type. The coleoptiles of this mutant showed 30–40 per cent less lateral (downward) redistribution of the plastids in response to geotropic stimulation than the wild type, lateral transport of auxin was 40–80 per cent less, and geotropic curvature was significantly diminished.[246] It would be interesting to carry out experiments on the roots of this mutant. The observations on coleoptiles are also considered to support the hypothesis that amyloplasts serve as gravity sensors. However, not all current workers consider that this view provides an adequate explanation of the phenomena.[15]

These observations and experiments afford elegant confirmation, by the use of modern techniques, of some of the classical theories. Charles Darwin demonstrated long ago that roots from which the terminal 1 mm or 0·5 mm had been excised were no longer sensitive to gravity. Early in the present century botanists held that certain solid cell inclusions or **statoliths**, principally starch grains, could transmit gravitational stimuli to the plasmalemma (ectoplast). Work with the light microscope demonstrated that starch grains did indeed collect at the lower sides of the cells of the root cap, and it was further shown that this did not occur in the root caps of roots which were not sensitive to gravity, from which starch grains were often completely absent.[231] This is thus another example where work with modern techniques and equipment confirms the shrewdness and ingenuity of much earlier workers, and it emphasizes again the overriding importance of ideas and hypotheses in science.

Observations with the electron microscope on the numbers of organelles in differentiating cells of the root cap indicate that the numbers of mitochondria, dictyosomes and the amount of endoplasmic reticulum are correlated with cell size, remaining fairly constant on a unit volume basis and increasing about 15-fold on a per cell basis. The formation of plastids, on the other hand, seems to be more or less restricted to dividing cells, and their number increases only about 3-fold, though their volume increases considerably.[297]

Epidermis

In the majority of roots, root hairs develop from some of the epidermal cells at a little distance from the apical meristem. If the root apex is damaged, root hairs may develop close to the tip of the root, suggesting that their formation is normally inhibited by it; however, other explanations are possible. Root hairs are outgrowths of single cells, and function both in the absorption of solutes and in anchorage. In some species the outgrowth which constitutes the root hair always emerges at or near the apical end of the cell, but its position at maturity depends on the amount of subsequent longitudinal growth of the epidermal cell. Root hairs are usually eventually sloughed off, but occasionally they are persistent. Recent work with the electron microscope indicates that the wall of the emerging root hair is a continuation of only the inner component of the wall of the epidermal cell which gives rise to it.[329]

In some species, root hairs are formed from special cells which are distinct in size and metabolism from the neighbouring epidermal cells. Such cells are known as *trichoblasts*. Some discussion of these interesting cells is given in Part 1,[127] Chapter 7. In the roots of the grass *Phleum*, trichoblasts are formed by unequal divisions of a cell of the immature epidermis, the *protoderm*. The trichoblast is the more apically (distally) situated cell of this division. In more recent work with the aquatic floating plant *Hydrocharis*, it was found that similar unequal divisions took place in the protoderm, but that the trichoblast was the proximal product of the division. Thus a small, densely cytoplasmic cell, the trichoblast, was formed at the proximal end of the original mother cell, and a larger, more vacuolate cell at the distal end (i.e. towards the root apex) (Fig. 2.9a).[128] These observations were confirmed with the electron microscope (Fig. 2.10). Cells in the

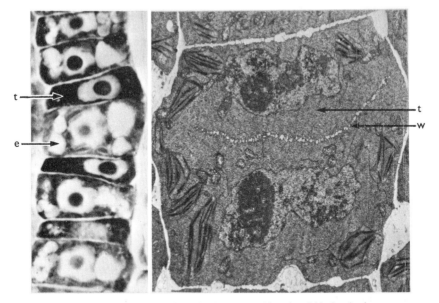

Fig. 2.9 Formation of trichoblasts in the root epidermis of *Hydrocharis morsus-ranae*. (a) Light micrograph of part of the protoderm showing the unequal division of the cells. The smaller proximal product of this division, which stains much more densely, is the trichoblast (t); the larger product is an epidermal cell (e). × 1200. (From Cutter and Feldman,[128] Fig. 4, p. 193.) (b) Electron micrograph of a protodermal cell in telophase stage of the unequal division giving rise to a trichoblast (t). The wall (w) between the two daughter cells, which is not yet completely formed, is markedly curved, a typical feature of unequal mitoses of this type. × 6000.

process of dividing unequally can sometimes be observed; the nucleus is displaced to the proximal end of the cell, and small vacuoles are usually present at the other end. The cell wall between the smaller and larger cells is typically curved (Fig. 2.9b). The unequal division and the appearance of the resulting daughter cells closely resemble those associated with the formation of a guard cell mother cell in leaves and stems of monocotyledons, except that there are less evident cytoplasmic differences between the two daughter cells (compare Figs. 2.9b and 2.10 with Fig. 5.29). Unequal divisions of this kind are important, since they are commonly the prelude to cell differentiation (see Part 1,[127] Chapter 2).

In *Hydrocharis*, the trichoblasts are very interesting cells. They do not divide, in contrast to the neighbouring larger cell which divides several times, but their nuclei continue to synthesize DNA. This results in their

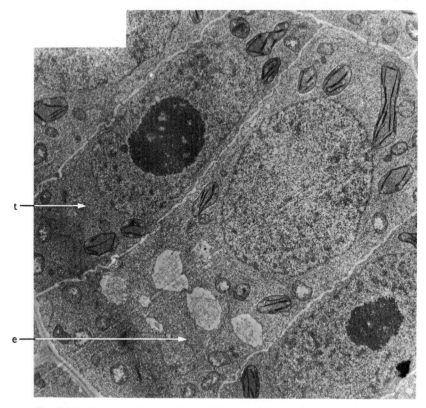

Fig. 2.10 Electron micrograph of a trichoblast (t) and adjacent epidermal cell (e) soon after the unequal division of a protodermal cell in the root of *Hydrocharis morsus-ranae*. The epidermal cell is more vacuolate than the trichoblast. × 6120.

becoming endopolyploid, with much larger nuclei than adjacent cells. Polyploid nuclei have multiples of the normal diploid amount of DNA. At some distance from the root tip the trichoblast nuclei have 8 times as much DNA as adjacent cells. This continued synthesis of DNA was demonstrated both by measuring the light absorption of nuclei stained with the Feulgen reagent with a microspectrophotometer, which enables comparisons to be made between nuclei, and by supplying the roots with ³H-thymidine.[129] In autoradiographs many more silver grains were present over the trichoblast nuclei. It is probable that the polyploid nature of the cell does not directly control its differentiation, but is a factor leading to its considerable growth. Root hairs in this species are especially large cells, about 5 mm long.

In other species, root hairs may develop from a whole row of cells, which

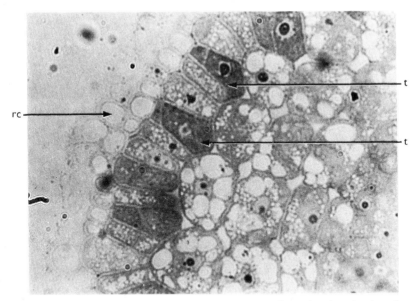

rc

t

t

Fig. 2.11 T.S. part of a root of radish, *Raphanus sativus*, showing the densely stained trichoblasts, which occur in longitudinal rows. rc, root cap; t, trichoblast. × 500. (By courtesy of M. J. York.)

stain more densely than their neighbours (Fig. 2.11), or from less differentiated cells.

In the **aerial roots** of certain epiphytic orchids a multiple epidermis, or **velamen**, is present. This tissue is derived from the epidermal initials and may be several cells thick. The cells are dead and devoid of contents; the cell walls are strengthened with bands of lignin. On the inner side of the velamen there is a specialized layer of cells which is derived from the periblem and not the dermatogen, and may therefore be considered as the outermost layer of the cortex, the **exodermis** (Fig. 2.12). This layer is composed of alternating long and short cells; the long cells become thick-walled on their radial and outer tangential surfaces, but the small cells remain thin-walled and are called **passage cells**. In some species, the cells of the velamen abutting on the passage cells are specialized in various ways; the causes of this might be worth investigating. The exodermis thus occupies a limiting position between the thin-walled cells of the cortex on the one hand, and the thick-walled cells of the velamen on the other; in this, as in the tissue of its origin, it is comparable with the endodermis (see below), and indeed the endodermis and exodermis are often mirror images on opposite sides of the cortex.[533] It is noteworthy that in both these tissues

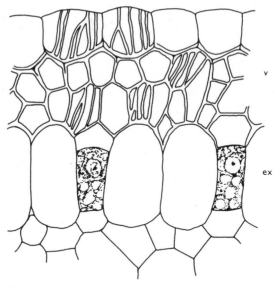

Fig. 2.12   T.S. part of the velamen (v) and exodermis (ex) of the aerial root of an orchid. The cells of the velamen have thickenings on the walls. The smaller passage cells of the exodermis are shown with cell contents. (From Fritsch, F. E. and Salisbury, E. J. (1961), *Plant Form and Function*, Bell and Sons, London, Fig. 64, p. 112.)

the tangential walls resemble the tissues on which they abut; i.e. the endodermal cells are thickened on the inner tangential wall, those of the exodermis on the outer wall. The reasons for this are not known, but it may perhaps be a consequence of purely mechanical factors.

The velamen is believed to function as a protective tissue, preventing undue water loss from the delicate cortical cells of the exposed aerial root. However, the fact that the apical tissues of the root are scarcely at all protected by the mature velamen, although equally exposed to desiccation, suggests that this conclusion might bear re-examination. It was thought formerly that the cells of the velamen also absorbed and conserved water drawn from the atmosphere, but more recent experiments indicate that the mature velamen and exodermis are nearly impermeable to water and solutes.[152]

Cortex

In most roots the cortex is parenchymatous. During development, the size of the differentiating cortical cells increases considerably before vacuolation becomes evident. In some roots, notably those of some aquatic plants, the cells of the cortex are very regularly arranged, both radially and in concentric circles. Conspicuous intercellular spaces may be present, and especially evident in aquatic species, where they form a type of aerenchyma. The cortical cells often contain starch, and sometimes crystals. Sclerenchyma is more common in the roots of monocotyledons than in those of dicotyledons. The characteristic trichosclereids found in the roots of *Monstera* develop from small, richly protoplasmic cells which are formed at the end of a file of cortical cells.[46] Collenchyma is occasionally present in roots; *Monstera* again provides a good example. The outermost layer or layers of the cortex, just beneath the epidermis, may be differentiated as an ***exodermis***, a kind of ***hypodermis***, with suberized walls. The innermost layer of the cortex is usually differentiated as an ***endodermis***. These layers have been shown to be histochemically somewhat similar.[533]

Endodermis

The differentiation of the endodermis is of especial interest since it comprises a single layer of cells differing physiologically and in structure and function from those on either side of it. Van Fleet[534] and others have shown that the enzymes peroxidase, cytochrome oxidase and polyphenol oxidase, and several others, are detectable in the developing endodermis; these enzymes are not restricted to the cells of the endodermis, of course, but must presumably occur there in different concentrations or in different sites in the cell. Van Fleet claims that histochemical methods reveal at least 14 different types of endodermis. In the young endodermal cells a band of suberin, the ***Casparian strip***, runs radially around the cell and is thus seen in the radial walls in transverse sections of roots (Fig. 2.13). This suberin deposit, to which the protoplast of the cell is attached, is continuous

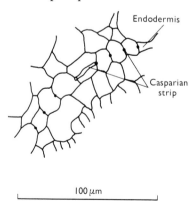

Endodermis

Casparian
strip

100 μm

Fig. 2.13   Part of T.S. young root of *Cucurbita maxima*, showing the developing endodermis with Casparian strips.

2

across the middle lamella of the radial walls, but is absent from the tangential walls. The Casparian strip is thought to exert control over the movement of materials in the root, but the function of the endodermis is still somewhat obscure. Possibly substances whose distribution is limited by the endodermis may include natural auxins. In some roots the cells of the exodermis also have Casparian strips.[533] A recent study of the endodermis with the electron microscope[50] reveals a thickening of the wall in the region of the Casparian strip. The plasmalemma is thicker here and adheres strongly to the cell wall. While plasmodesmata are present between endodermal, cortical and pericyclic cells, none was observed in the region of the Casparian strip.

In roots which do not undergo secondary thickening a suberin lamella

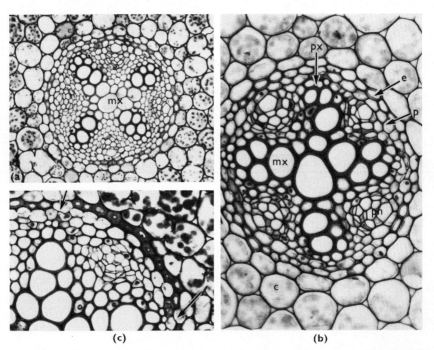

(c)   (b)

Fig. 2.14   T.S. root of *Ranunculus*. (a) Young root, showing the tetrarch vascular cylinder surrounded by parenchymatous cells containing starch grains. The metaxylem (mx) is not yet fully differentiated.   × 50. (b) Slightly older root in which the vascular tissue is fully differentiated. Only part of the endodermis (corresponding in position to the phloem) is thick-walled. c, cortex; e, endodermis; mx, metaxylem; p, pericycle; ph, phloem; px, protoxylem.   × 250. (c) Still older root, in which the endodermal cells are thick-walled except for unthickened passage cells (arrowed) outside the protoxylem poles.   × 250.

usually forms over the whole inner wall of the endodermal cells; later, cellulosic thickening which often becomes lignified is deposited on the inner tangential and the radial walls of the cells (Fig. 2.14b). The thick-walled nature of the endodermis is also evident in studies with the electron microscope. Formation of the wall thickening usually begins in the cells of the endodermis in positions adjacent to the phloem and spreads radially from there, so that thin-walled *passage cells* often remain in the endodermis in positions opposite the protoxylem (Fig. 2.14c). The nature of this localized stimulus to the formation of wall thickening has not been examined in any detail.

Pericycle

The pericycle is usually a single layer of parenchymatous cells lying just within the endodermis and peripheral to the vascular tissues (Fig. 2.14b) and derived from the same initials in the meristem. It retains a capacity for meristematic growth, and gives rise to lateral root primordia, parts of the vascular cambium (a lateral meristem which gives rise to secondary xylem and phloem) and usually also the meristem which produces cork, the phellogen. The cells of the pericycle, like those of the apical meristem, are normally always diploid, although other neighbouring tissues may show various degrees of polyploidy.[521] The pericycle is sometimes called pericambium.

Vascular tissues

The vascular system of the root as seen in cross section consists of a variable number of triangular rays of thick-walled, lignified tracheary elements, alternating with arcs of thin-walled phloem (Fig. 2.14a, b). In the root, in contrast to the stem, the *xylem* and *phloem* do not lie on the same radius. The xylem may form a solid central core, or there may be a parenchymatous or sclerenchymatous *pith*, as in the roots of many monocotyledons. Roots with 1, 2, 3, 4, 5 and many arcs of xylem are respectively called monarch, diarch, triarch, tetrarch, pentarch and polyarch.

In the root, the direction of differentiation of the *procambium* is from older, more mature tissue towards the root apex, i.e. acropetally. The procambium consists of densely staining meristematic cells which are elongated in the longitudinal plane of the organ in which they occur. From these cells mature xylem and phloem eventually differentiate. The whole of the central cylinder is usually regarded as procambium, whether or not a central pith eventually differentiates. Alternatively, the region which will differentiate as pith may be called *ground meristem*.[168] Differentiation of the vascular elements can be studied appropriately in transverse sections taken at various levels proximal to the root apex, i.e. proceeding towards the base of the root (basipetally). Such sections enable us to study differentiating cells both with respect to space and to time. Since the root is continuously growing acropetally, the tissues observed in a section at a particular level behind the tip would have changed and further differentiated if left to grow for a further period of time. Progressive tissue differentiation can be studied also in longitudinal sections (Fig. 2.15).

The first vascular elements to become distinguishable in sections taken behind the root apex are the immature **metaxylem** vessel elements. These

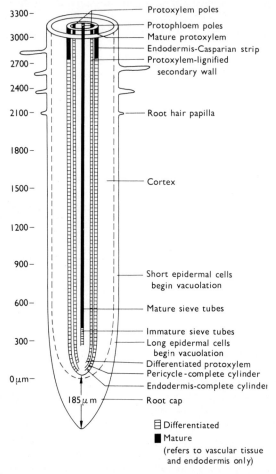

Fig. 2.15 Diagram of a root tip of *Sinapis alba* illustrating levels of differentiation and maturation of various tissues proximal to the root apex. The levels shown for each tissue are the averages of measurements made on sections of 15–20 roots of seedlings grown on moist filter paper. (From Peterson,[399] Fig. 10, p. 326. Reproduced by permission of the National Research Council of Canada.)

(a)

(c)

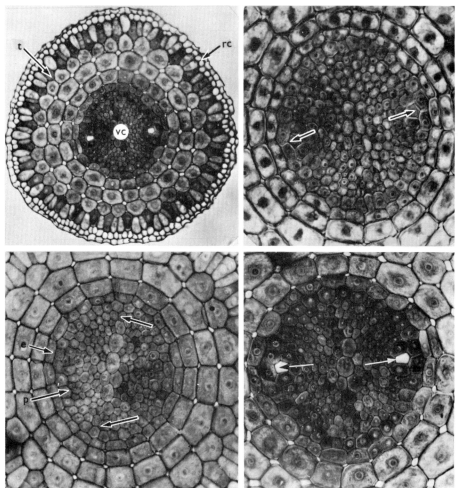

(b)

Fig. 2.16 Tissue differentiation as seen in transverse sections of the root of **(d)** *Sinapis alba*. (a) At about 450 μm from the root apex two sieve tubes (angular cells without visible contents) are differentiated in the vascular cylinder (vc). Two layers of root cap (rc) cover the epidermis, which consists of trichoblasts (t) and ordinary epidermal cells. The cortex (c) consists of regularly arranged cells with intercellular spaces. × 150. (b) Section 60 μm from the apex. Arrows indicate the outermost, still differentiating protoxylem elements of the diarch xylem plate. e, endodermis; p, pericycle. × 300. (c) Approximately 180 μm from the root apex. Arrows indicate differentiating sieve elements. × 300. (d) Transverse section of the vascular cylinder approximately 450 μm from the root apex. Arrows indicate two fully differentiated sieve elements of the protophloem. × 300. (From Peterson,[399] Figs. 1, 4, 5 and 7. Reproduced by permission of the National Research Council of Canada.)

appear as large, somewhat vacuolate cells in T.S. (Fig. 2.16b). At a slightly more proximal level in the root the first sieve tubes of the *protophloem* differentiate, appearing in T.S. as angular cells which are eventually devoid of contents (Fig. 2.16a, d). More proximally along the root the first lignified elements of the *protoxylem* become distinguishable at the periphery of . the xylem rays; the future metaxylem elements, although clearly distinguishable, are still immature and unlignified at this level in the root. The xylem in the angiosperm root is *exarch*, i.e. it differentiates centripetally from the periphery of the vascular cylinder towards the centre (Fig. 2.16b), as opposed to the *endarch* or centrifugal xylem of angiosperm stems. In the roots of some monocotyledons a single metaxylem element is present in a central position; in others numerous metaxylem elements surround a central pith. Differentiation of the xylem and phloem, as of the procambium, is acropetal.[165]

The distance behind the root apex at which mature elements of the protophloem and protoxylem differentiate varies according to the rate of growth of the root, and to other conditions. In roots maintained artificially in aseptic culture media this distance has been shown to be affected by the degree of aeration of the medium, by the sugar concentration, and by other factors,[496] all of which presumably in turn affect rates of growth. In general, tissues become differentiated closer to the root apex in more slowly growing roots. In some roots the levels of initiation and of final differentiation, or maturation, of tissues appear unrelated to one another;[428] levels of maturation are affected by rate of growth very much more than levels of the first discernible differentiation.[245] However, in some species, at least, it can be shown that differentiation of all the various tissues is affected by rate of growth of the root, and that, in general, their maturation still takes place in the same order.[399, 409]

The first xylem of the root often becomes mature proximal to the region of elongation, and, as a consequence of this, tracheary elements with annular and helical (spiral) types of secondary wall thickening may not occur in the protoxylem.[168]

Histochemical studies have shown that peroxidase activity is present especially in the root cap, the differentiating epidermis and the early vascular tissue. After treatment of the roots with indoleacetic acid (IAA) increased formation of peroxidase was induced in the cells of the vascular tissue.[276] This observation may be of some value in interpreting the numerous effects of IAA on root tissues now reported, though the function of peroxidase itself is by no means clear.

Factors controlling xylem differentiation in roots

In recent ingenious experiments with pea roots, Sachs[440-442] has shown that auxin transported from the shoot may be important in affecting vascular differentiation in roots. These experiments involve the differentiation of cortical parenchyma cells as tracheary elements under the influence of a source of auxin. (Experiments of this type, carried out on shoots, were discussed in Part 1,[127] Chapter 8.) In the first of his experiments, Sachs[440] severed the root from the shoot of pea seedlings and grafted the two back together again. A connection between the xylem on both sides of the cut was formed in a few days by the differentiation of parenchyma cells as xylem cells. If the grafts were prevented from touching, xylem differentiated only on the shoot side of the cut. Since the influence of the shoot could be replaced by 1 per cent IAA in lanolin, it is concluded that the shoot normally supplies a xylem-inducing stimulus, probably auxin, to the root. Sachs suggests that the root essentially acts as a sink for a stimulus from the shoot, and that xylem is formed along the path of the stimulus, much as it is known to do in wounded stems (see Part 1).

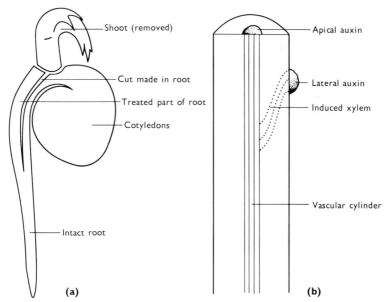

Fig. 2.17 The system used for experiments on the induction of xylem in pea roots. (a) 3-day-old pea seedling, showing the cuts made to separate the half-root used for the experiments. It remained connected to the cotyledons through the intact distal part of the root. (b) Cut surface of the treated part of the root. The two places to which auxin was applied are indicated. (After Sachs,[441] Fig. 1, p. 785.)

In other experiments Sachs[441] partially split roots of decapitated pea seedlings longitudinally in such a way that contact between the two halves was maintained through the root tip (Fig. 2.17). If IAA in lanolin was applied to the side of the half root, cortical cells differentiated as xylem and the strand of xylem formed a connection with the pre-existing root vascular tissue. If IAA is applied to the root not only laterally, but also at the severed end of the central cylinder, the induced strand of xylem does not join the central cylinder. Sachs[441] concludes that a xylem strand will develop towards pre-existing vascular tissue only if this is devoid of a source of auxin. Under such conditions he considers that the tissue acts as an attractive sink for a new source of auxin. On the contrary, xylem strands are inhibited from forming in the vicinity of vascular tissue that is well supplied with auxin.

Refining these experiments further, Sachs[442] used half-roots, as in the previous experiment, but removed the existing central cylinder of vascular tissue. Under such conditions, IAA applied laterally induced xylem strands which connected with vascular tissue in the intact part of the root tip. If two lots of auxin were applied in different sites, one application preceding the other by a week, the newer strand united with the first one. Thus, in a tissue devoid of normal vascular tissue, cells which have begun to differentiate as vascular elements attract the inducing stimulus—auxin—and prevent the differentiation of neighbouring cells as xylem elements.

These findings led Sachs[440] to conclude that a xylem-inducing stimulus originating in the shoot moves downwards to the roots through the vascular system, and that at least under experimental conditions xylem is formed along the path of the stimulus. Presumably, therefore, such a factor could also be operative in the normal differentiation of xylem in the root. However, the experiments of Torrey[517] on cultured pea root tips, discussed below, seem to indicate that an influence from the shoot is not essential for vascular differentiation.

Patterns of vascular differentiation

The beautifully regular pattern of the vascular tissues is one of the most characteristic features of roots, and is a good example of the symmetrical arrangement of many plant organs and tissues. What factors control these regular and characteristic patterns? In 1952 Turing[532] put forward a theory which attempted to explain, in mathematical terms, how a pattern could arise in an initially homogeneous system. In its simplest form this explained how two morphogenetic substances and their associated metabolic materials could come to be distributed or aggregated in an invisible pattern described as a 'stationary wave'. This distribution of the substances could be the basis for the formation of a visible pattern of organs or tissues.

That this is mathematically possible is not in doubt, but it has not yet been demonstrated biologically; indeed the techniques for this are probably not yet available. Bünning[71] has suggested that vascular pattern may result from forces which induce the differentiation of xylem in proximity to already differentiated xylem elements, in radial and longitudinal planes, and inhibit the differentiation of phloem, which therefore appears as far as possible from the xylem groups (differentiation of immature metaxylem having established these first). Thimann[505] has pointed out that it is unlikely that diffusible substances such as auxins or vitamins could be operative in systems of this kind, since the necessary sharp concentration gradients would not occur. Compounds of high molecular weight would be more likely to be involved. Bünning's hypothesis[71] is further open to criticism on other grounds.[168]

Yet there is experimental evidence that hormones of the auxin type are somehow involved in the establishment of vascular pattern, though possibly indirectly, through their effects on the apical meristem.

Although it is apparently not true of all species, it has long been known that in many plants there is a general relationship between the actual number of rays of xylem and arcs of phloem and the diameter of the root.[545] Subsequently it was suggested[506] that the factors controlling this phenomenon should be sought in the root apical meristem, where the tissues originated. Some recent experiments on the aseptic culture of excised roots substantiate this view.

Torrey[517] was able to grow pea roots in artificial nutrient media from the terminal $\frac{1}{2}$ mm of the root tip, a fact of considerable interest in itself, since much of this region is occupied by the tissues of the root cap and the quiescent centre, which usually divide relatively slowly. At the base of these $\frac{1}{2}$ mm tips the triarch vascular system characteristic of the pea root was just becoming established, but no differentiated vascular elements were present. In some of these root tips grown in culture, there was a temporary modification of the vascular system to the monarch or diarch condition.

In experiments with bisected root tips of *Sinapis*, Reinhard[422] also showed that the number of xylem groups could be modified. New apices regenerated from the two halves, and formed new roots. Sometimes both of these were triarch, instead of the normal diarch condition, and sometimes one half was diarch and the other triarch.

These experiments established two important points. (1) As observational studies already indicated, the pattern formed by the developing vascular tissues is evidently due to the activity of the apical meristem, and does not result from inductive influences from the older vascular tissues, even though the actual differentiation of the tissues is acropetal. (2) In Torrey's experiments, the radial arrangement of the primary vascular tissues, i.e. the number of strands of xylem and phloem, appeared to be related to the dimensions of the apical meristem at the level at which the inception of the pattern occurs.

The first of these conclusions is possibly also supported by some experiments carried out by Bünning.[70, 71] The apical 2 mm of roots was excised,

turned through an angle, and replaced on the remaining stump of older
root tissue. The vascular tissues which subsequently developed in these
root tips were out of line with those in the older parts of the roots. Thus
here also the vascular pattern was apparently determined in the tip and not
by the existing older vascular tissues. However, as Esau[168] has rightly
pointed out, some differentiation of vascular tissues would have already
occurred in tips 2 mm long.

In another experiment with cultured roots, the developing vascular
pattern was changed in a different way. Torrey[518] found that the new apical
meristems which regenerated on decapitated pea roots in a medium containing appropriate concentrations of IAA had a symmetrical hexarch
vascular system, i.e. one with 6 strands of xylem and phloem, instead of the
3 strands characteristic of normal pea roots and of those formed in control
medium. The hexarch vascular system was maintained as long as the root
was kept in a medium containing IAA, but if the root was transferred to
control medium without IAA it eventually reverted to a tetrarch or triarch
pattern. After transference to control medium the hexarch root developed
symmetrical pentarch and tetrarch vascular patterns before attaining the
typical triarch state.

These findings are compatible with the previous conclusions, because
the diameter of the procambial cylinder of hexarch roots at the level of
pattern formation was greater than that of triarch roots. It is suggested that
the artificially supplied auxin (IAA) has the effect of increasing the size of
this region of the root meristem. Possibly this increase in dimension of the
pattern-forming region was sufficient to minimize the inhibitory influence
of one vascular strand on another demonstrated by Sachs.[441] It is only a
further logical step to postulate that under normal conditions of growth the
dimensions of the root meristem may be controlled by the endogenous
auxin in the root, i.e. that naturally present in its tissues.

In further experiments with regenerating root tips, Torrey[520] has shown
that if, in addition to IAA, kinetin is supplied at certain concentrations root
elongation is inhibited and a quite different, concentric vascular pattern,
resembling that of secondary vascular tissues, is induced. This is apparently
a consequence of an increased capacity of certain cells to divide.

These experiments thus indicate that hormonal substances, alone or in
various combinations, are probably important in controlling the site and
amount of cell division in the root apical meristem, and that this in turn
can affect vascular pattern. It then becomes important to discover the
site of production of endogenous hormones in the root. It is considered
possible that the terminal region of the meristem, especially the quiescent
centre, may be one site of production of the substances involved.[520] We
have already seen that the cells of the root cap apparently exert some control
over the distribution of the hormonal substances involved in geotropism.

Perhaps the effect of the root cap on the control of vascular pattern should also now be investigated.

These experimental results offer some explanation of the factors controlling the number of strands of xylem and phloem in the root, but do not explain why some procambial cells differentiate as elements of xylem and others as components of the phloem. Experiments with stem callus tissue have indicated that the balance between auxin and sugar concentrations is important in controlling the differentiation of these elements (see Part 1, Chapters 8 and 9), and it seems likely that similar factors must be operative in roots, where, however, xylem and phloem are not radially aligned but alternate with one another during primary growth. We may now be close to achieving some understanding of the factors controlling vascular differentiation, as a result of a combination of careful observational work and experimental investigations. Much more information is required, however, about the sites of synthesis of the substances involved and about the mechanism of their distribution (perhaps in 'stationary waves') within the root.

ORIGIN OF LATERAL ROOTS

In flowering plants, the primordia of lateral roots are usually formed in the pericycle, although in some species the endodermis also participates in their formation. In ferns and other pteridophytes lateral roots originate in the endodermis. The primordia of lateral roots thus have a deep-seated, or *endogenous*, origin, having to emerge to the exterior by growing through many layers of tissue; this contrasts with the relatively superficial, or *exogenous*, origin of the leaf and the majority of bud primordia (see Chapter 3). In the inception of a lateral root, the cells of a localized region of the pericycle some distance behind the root tip divide periclinally, i.e. by walls formed parallel to the surface of the root. These first divisions are followed by others in both the periclinal and anticlinal (at right angles to the root surface) planes (Figs. 2.18, 2.19a, b). The cells of the endodermis may also divide. The small group of meristematic cells thus formed gradually becomes organized into an apical meristem resembling that of the parent root, and pushes its way through the cortex and epidermis (Fig. 2.19c). The question of whether enzymatic processes are involved is still somewhat controversial, but work with the electron microscope[51] indicates that the protoplasm of adjacent cortical cells is destroyed by the growing root. The walls of these cells remain and envelop the growing lateral root primordium. A study of the distribution of the enzyme β-glycerophosphatase in pea roots has shown that it is generally associated with cells about to die and undergo lysis, including cortical cells surrounding emerging lateral roots.[464] It is suggested that the specialized coated vesicles observed

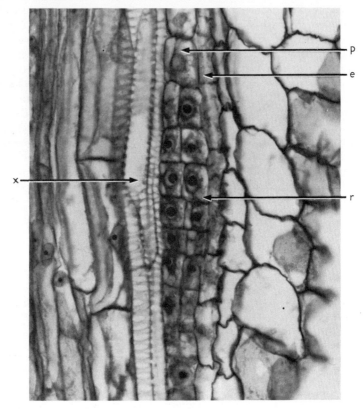

Fig. 2.18 L.S. root of safflower, *Carthamus tinctorius*, showing the origin of a lateral root primordium (r) by division of the cells of the pericycle (p). e, endodermis; x, xylem.   × 600. (By courtesy of Dr. T. E. Mallory.)

near dictyosomes in the outer layer of cells in the root primordium may transport hydrolytic enzymes to the cell wall;[51] thence they would presumably move out and attack adjacent cortical cells. This is an interesting topic for further work, especially since these observations apparently apply to root primordia but not to buds which originate endogenously. Eventually vascular tissues are differentiated in the lateral root primordia, and a connection is established between elements of the xylem and phloem of the main and lateral roots.

The lateral root primordia are formed in distinctive positions in relation to the xylem and phloem of the parent root. In diarch roots they usually occur between the xylem and phloem, in triarch and tetrarch roots in positions opposite to the protoxylem (Fig. 2.19c) and in many polyarch

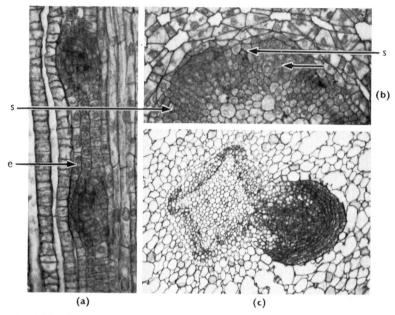

Fig. 2.19 Origin of lateral root primordia. **(a)** L.S. root of *Victoria trickeri*, showing two lateral root primordia which have resulted from division in the peri-cycle and underlying cell layers. The endodermis (e) is probably also dividing. Conspicuous air spaces are present in the cortex.   ×250. **(b)** T.S. root of *V. trickeri*, showing origin of lateral root primordia between sieve elements (s) of the protophloem, i.e. opposite the protoxylem. The layer of cells below the pericycle has divided periclinally (arrow) in early stages of root formation.   ×250. **(c)** T.S. root of *Salix*, showing a well-organized lateral root primordium growing through the cortex of the parent root.   ×100.

roots opposite to the protophloem. In some polyarch roots, however, the lateral roots are formed in sites opposite to the protoxylem.

In the roots of some species, a number of which are aquatic plants,[17] lateral root primordia are formed close to the root apex, in still meri-stematic regions. In such roots lateral root primordia are numerous and are formed in close proximity and this raises the question of whether they are arranged in regular patterns comparable with those formed by leaf pri-mordia in the shoot. Recently some preliminary information on this point has become available. Working with roots of banana (*Musa acuminata*) and five other monocotyledons, in which primordia are not formed especially close to the meristem, Riopel[426, 427] analyzed mathematically the nearest neighbour distances for lateral root primordia. His results indicated that lateral root primordia were not randomly arranged, but occurred in a complex non-random, dispersed distribution. Certain of the sites opposite the protoxylem are apparently more favourable for root initiation than others, on the basis of the position of the previous primordium. Other work[350] has shown that in species with many root primordia close to the meristem, primordia are sometimes formed in adjacent sectors of the root, i.e. opposite to adjacent protoxylem or protophloem arcs, at approximately the same level. It therefore seems unlikely that there can be much inhibition between primordia in the horizontal plane, at least in these species. In the longitudinal plane, however, lateral root primordia are found to be approximately equally spaced in the various sectors of the same root; this spacing varies somewhat in different roots of the same plant.[350] This relationship suggests that the position of a lateral root primordium is affected both by existing lateral roots in the same sector, and by the apical meristem itself, since primordia in the various sectors are formed at a fairly constant distance from the tip of the root. In the species studied, the relationship was precise enough to permit prediction of the site of the next primordium.

This effect of the apical meristem has also been observed by others,[519] who have shown that branching of the root is stimulated by its decapitation.[521] Some of the work on the factors affecting lateral root formation fails to make the important distinction between effects on inception or on outgrowth of primordia, and is accordingly difficult to interpret.

It appears that in many roots auxin is one of the essential factors for lateral root formation, and that it specifically stimulates cell division in certain regions of the pericycle.[521] In pea roots, there is evidence that a factor other than auxin, promoting the formation of lateral roots, is present in the older basal regions of the roots and is translocated towards the root apex.[396, 397]

Other recent experiments suggest that determination of the sites of lateral root primordia, and early stages in their development, can take place without the final stages of nuclear and cell division. When wheat seedlings were treated with colchicine, which permits synthesis of DNA but prevents the later stages of mitosis, regions shaped like root primordia, called **primordiomorphs**, were formed at normal positions in the pericycle. Their cells had nuclei larger than those of adjacent cells, but they did not divide. Upon removal of the colchicine, the primordiomorphs underwent cell division and developed into lateral roots.[179] Somewhat similar structures were induced in the roots of other species by trifluralin, a herbicide.[36] It seems likely that the initial trigger to cell division in regions of the pericycle is the stimulus to lateral root formation, subsequent stages following in an epigenetic fashion.

The existing evidence concerning the factors involved in lateral root formation again seems to implicate hormonal substances. This emphasizes the need to know more about their distribution in the root and their mode of action at the cellular level.

SECONDARY GROWTH

The roots of gymnosperms and most dicotyledons undergo secondary growth, during which a **vascular cambium** develops and gives rise to secondary xylem and phloem, and a **phellogen**, or cork cambium, is formed and produces the **periderm**. The roots of extant vascular crypto-gams and most monocotyledons do not undergo secondary growth.

The vascular cambium originates from procambial cells which remain undifferentiated between the primary phloem and the primary xylem. Thus at first it consists of separate bands of tissue. Subsequently the cells of the pericycle adjacent to the protoxylem poles also give rise to cambium, and this joins up with the existing strips to form a continuous tissue. At this stage the cambium is not cylindrical but lobed, following the outline of the primary xylem (Fig. 2.20a). The cambium which originated from the pro-cambial cells on the inner side of the protophloem becomes active first, however, and forms a considerable amount of **secondary xylem**, until the cambium appears circular in T.S. (Fig. 2.20b). The cambium divides peri-clinally, producing **secondary phloem** towards the outside of the root and secondary xylem towards the inside. It thus functions in a similar manner to the cambium in the stem, and forms cylinders of secondary xylem and phloem. Indeed, the structure of an old root with a considerable amount of secondary tissue superficially resembles that of a secondary stem, except that in the primary xylem the exarch protoxylem typical of roots can usually still be distinguished (Fig. 2.20c). In some roots, the cam-bium cells derived from the pericycle may give rise to parenchymatous ray tissue. When considerable secondary growth takes place in the primary phloem, endodermis, cortex and epidermis may be sloughed off.

After the production of secondary vascular tissues has begun in the root, the formation of periderm usually ensues. The cells of the pericycle divide actively, and the **phellogen** or cork cambium arises in the outer layers. This consists of a cylinder of tissue which functions in a manner comparable with the vascular cambium, forming **phellem** or cork towards the outside of the root and, rather later, **phelloderm** or secondary cortex centripetally. The cells of these tissues are consequently radially aligned with the cells of the phellogen, as in the stem (see Chapter 4). The tissues outside the developing phellem usually become sloughed off.

Why the remaining procambial cells in a growing root should suddenly become activated and function as an active meristematic tissue, the vascular cambium, giving rise to secondary xylem and phloem, is an interesting and important question. It is known that, in stems of plants in the North

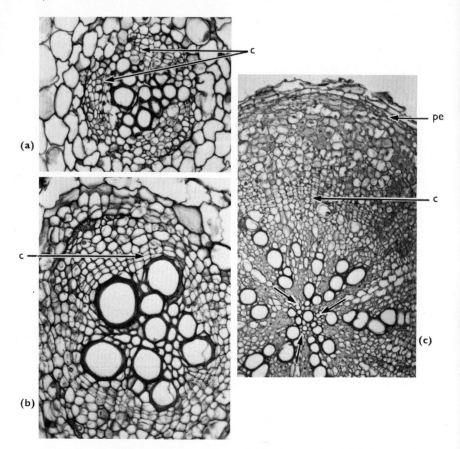

Fig. 2.20 Secondary growth in the root of *Medicago sativa*. (a) Early stage in secondary growth, with the cambium (c) still following the triangular outline of the triarch primary xylem.   × 250. (b) Later stage, with the cambium (c) almost circular in T.S. Some secondary xylem and phloem have been formed.   × 250. (c) T.S. older root, in which considerable amounts of secondary xylem and phloem have been formed from the cambium (c). The original protoxylem poles (arrowed) are still distinguishable. Xylem peripheral to this is secondary. A periderm (pe) is present at the exterior, the cortex and epidermis having been sloughed off.   × 100.

Temperate region, the cambium becomes activated at the time of bud growth in the spring, stimulated by the basipetally transmitted auxin produced by the growing buds (see Chapter 4). Until recently, little was known of the causal factors stimulating cambial activity in roots, although Garner and Allard[197] had shown that secondary thickening was dependent on daylength. In 1963, however, Torrey[520] extended his experimental work with excised pea roots to include a study of the factors controlling the development of the vascular cambium and secondary vascular tissues. He found that 'initial tips', i.e. tips excised from seedling roots, elongated considerably when grown in the control nutrient medium, but formed no cambium. In control medium + IAA the elongation was inhibited, sometimes by as much as 90 per cent, lateral roots were formed and vascular cambium developed in basal parts of the roots. 'First transfer tips', i.e. tips excised from the cultured roots and transferred to fresh medium, did not form cambium even in the presence of IAA. Torrey concluded that 'initial tips', when stimulated by auxin, were rich in substances essential for the inception of the vascular cambium, but that this substance, or these substances, were used up or diluted in tips which had been cultured on control medium for a week. He concluded that in the 'initial tips' cambium was formed in response to a hormonal stimulus moving acropetally, i.e. towards the root apex, over a period of time.

In another series of experiments Torrey[520] tried to supply 'first transfer tips' with various substances from the base of the excised root tip, in an attempt to simulate conditions in the seedling, where the roots are continually receiving substances from the cotyledons and other aerial parts of the plant. A technique was devised for supplying the growing root tips with one medium through the base, contained in a small tube, and another through the absorbing surfaces of the root, contained in a Petri dish (Fig. 2.21). It was found that supplying sugar + IAA *via the basal surface* led to

First-transfer root tip

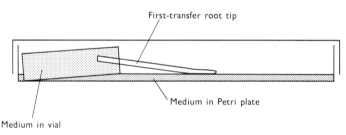

Medium in Petri plate

Medium in vial

Fig. 2.21 Method for supplying cultured root tips with one medium in the Petri dish and another, supplied via the basal end of the root, in a vial. (From Torrey,[520] Fig. 4, p. 300.)

the formation of vascular cambium at the base of the root and its acropetal differentiation within the growing root. The conditions found to be most successful for root elongation and extent of cambial differentiation were 8 per cent sucrose + IAA (at a concentration of $10^{-5}M$) + a mixture of additional factors, in the mineral nutrient medium supplied via the base of the root, and 4 per cent sucrose in the medium in the Petri dish. The *essential*

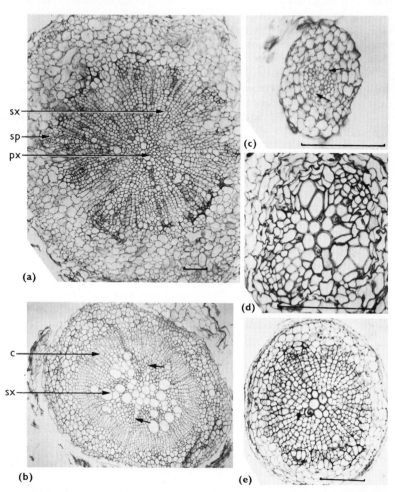

Fig. 2.22 Experimental stimulation of vascular cambium in the root of radish, *Raphanus sativus*. (a) T.S. root supplied with complete vial medium (modified Bonner medium (basal) $+8$ per cent sucrose $+10^{-5}$ M IAA $+5 \times 10^{-6}$ M benzyladenine $+5 \times 10^{-4}$ M myoinositol), fixed after 34 days in culture. Extensive secondary xylem (sx) and some secondary phloem (sp) have been formed. px, primary xylem. $\times 65$. (b) T.S. root grown for 10 days in a growth chamber with artificial light under a 16-hour photoperiod. The cambial region (c) is many-layered and a considerable quantity of secondary xylem (sx) has been formed. Protoxylem poles arrowed. $\times 60$. (c) T.S. root provided with vial medium without growth factors, fixed after 31 days in culture. No cambium or secondary tissues are present. Protoxylem poles arrowed. $\times 230$. (d) T.S. root supplied with basal

factor for the initiation of vascular cambium in pea roots, however, was the provision of auxin (IAA) via the base. Under natural conditions, this is presumably supplied by the aerial parts of the plant.

In subsequent experiments with excised radish roots,[335] it was shown that first transfer roots supplied with IAA and sucrose via the base of the root did not form cambium, or formed it in very limited amount. In these roots, other factors were necessary for the stimulation of cambium. It was found that in the presence of sucrose both auxin and a cytokinin were necessary in order to effect cambial stimulation (Fig. 2.22). Various cytokinins were effective, including 6-benzylaminopurine and kinetin itself.[522] The amount of secondary thickening was greatly promoted by myoinositol (Fig. 2.22).

These experiments indicate that substances affecting growth, either auxin alone or auxin together with a cytokinin, can stimulate cambium formation in excised roots, but only when supplied so that they are transported acropetally in the root. In the whole plant, the presumptive source of these substances—or at least of auxin—would be the aerial parts. Further experiments on pea roots support this hypothesis. Shoot or root apices were excised from pea seedlings 3 or 5 days after germination, thin slices were excised daily to give a fresh cut surface and lanolin paste containing IAA was applied to the cut surfaces. At intervals sections were made 2 mm below the insertion of the cotyledons, and the numbers of cells in the cambial region and of lignified xylem elements were counted. Excision of the root apex had little effect, but many fewer cells were formed in the cambial region following excision of the shoot apex.[142] Thus the shoot apex apparently stimulated cambium formation in the root, and this is apparently due to auxin moving down into the root, since applied IAA could compensate for the excised shoot apex.

These experiments emphasize the importance of various physiological interactions between the parts of the whole plant. In interpreting the structure of a particular organ, its relationships with the whole plant and the environmental conditions under which the plant has been grown should always be considered. As many examples testify, conditions of growth often affect structure.

---

medium + sucrose + benzyladenine + myoinositol, but lacking IAA, fixed after 29 days in culture. Limited secondary xylem was formed.   × 265. (e) T.S. root supplied with basal medium + sucrose + IAA + benzyladenine, but lacking myoinositol, fixed after 34 days in culture. Less secondary tissue has been formed, and the root is of smaller diameter; compare (a).   × 120. Lines indicate 100 μm. (From Torrey and Loomis,[522] Figs. 1–3 and 12, pp. 1100 and 1104, and Torrey and Loomis,[522a] Fig. 2, p. 404.)

Anomalous secondary structure

In some roots, secondary growth follows different courses from the so-called normal condition described, i.e. the one most commonly found. In many *storage roots*, e.g. carrot (*Daucus*), an unusually large amount of parenchyma is present in the xylem and phloem, but secondary growth is otherwise normal. In the beetroot (*Beta*), additional cambia are formed from the pericycle and phloem outside the normal vascular cylinder. These cambia are arranged concentrically, and each produces some phloem to the outside and xylem to the inside of the root; abundant storage parenchyma is also present. These concentric rings of cambium and secondary vascular tissues can be seen clearly in slices of beetroot, or of sugar beet, where they are less obscured by anthocyanin. In the roots of sweet potato (*Ipomoea batatas*), a cambium is present in the usual situation but, in addition, accessory cambia form around the tracheary elements.[244] These form some xylem elements towards the inside, i.e. towards the isolated groups of tracheary elements themselves, and some phloem towards the outer side.

It is evident that the physiological factors controlling cambium formation in these and other species with anomalous secondary structure—themselves obviously under genetic control—must be extremely complex, and much work remains to be done in this field.

3

The Stem: Primary Growth

The aerial parts of vascular plants consist of an axis, the stem, bearing lateral organs. The stem is usually upright or orthotropic, but may be decumbent, horizontal, or plagiotropic. During the vegetative phase the lateral organs which it bears are of two kinds: *leaves*, which are characterized by limited or determinate growth and are usually of dorsiventral symmetry though variable in shape, and *buds*, which are characterized by potentially unlimited or indeterminate growth and are usually of radial symmetry. The stem together with the leaves which it bears constitutes the *shoot*; the relationship between leaves and stem is a very close one and a separation of the shoot into its component parts is to some extent artificial. This distinction will, however, be made here, and a consideration of leaves is deferred until Chapter 5. In the discussion which follows, it should be borne in mind that comments may sometimes pertain to the leafy shoot rather than to the stem itself. Positions on the stem at which leaves occur are called *nodes*, the intervening leafless portions of the stem being *internodes*. The buds usually occur in the axils of the leaves, commonly one bud to each leaf axil, though a number of other arrangements also occur.

During the reproductive phase of development, *flowers* or *inflorescences* are borne either laterally or terminally on the stem, or in both positions. These develop either from lateral bud meristems, in which case they usually occupy axillary positions, or from the terminal bud itself. In this instance growth of the axis is terminated and can be continued only by a lateral bud, giving a *sympodial* shoot system. Where the same terminal vegetative bud continues to grow for several years, the shoot system is said to be *monopodial*; in such species flowers occupy lateral positions.

Growth of the stem is brought about by a number of **meristems**. The most important of these is the shoot **apical meristem**, the growing region of the terminal bud. As in the case of the root, all the primary tissues of the stem are derived from this meristem. Unlike the root apical meristem, it gives rise to lateral organs, forming leaf primordia and, in most instances, also the primordia of lateral buds. Terminally situated flowers or inflorescences develop from the apical meristem, which undergoes profound physiological and structural changes at the onset of the reproductive phase (see Chapter 6).

Leaf primordia are formed at the apical meristem in a regular sequence and pattern or phyllotaxis, that is usually characteristic of the species. The time interval between the formation of one leaf primordium and the next is known as a **plastochrone**. Various systems of **phyllotaxis** are recognized; the simplest types may be summarized here. The commonest type of leaf arrangement is probably **spiral** phyllotaxis, in which one leaf occurs at each node and a spiral, or more strictly a helix, can be drawn round the stem linking the leaves in the order of their formation. Successive leaf primordia are separated in the tangential plane by an angle of divergence; in Fibonacci spiral systems, this angle approaches the so-called **Fibonacci angle**, approximately 137·5°. If internodes are short, two series of helices can usually be recognized passing round the stem in opposite directions, linking up certain of the leaves. Even if these helices, the **parastichies**, cannot easily be seen on the mature stem, the most obvious of them, the **contact parastichies**, can usually be recognized in transverse sections of the shoot apex (Fig. 3.1). In most species (though not in ferns and a few angiosperms) the leaves along these helices, or parastichies, are in contact at their bases; hence the name 'contact parastichies'. The numbers of the contact parastichies in the two directions are always two consecutive terms from the **Fibonacci series**, a mathematical series in which each term is the sum of two preceding terms. The commonest Fibonacci series is 0, 1, 1, 2, 3, 5, 8, 13, 21, 34, 55, . . . and so on. Systems of spiral phyllotaxis can be described by the numbers of the two intersecting sets of contact parastichies, contained in parentheses and with a + sign between, i.e. (2+3), (3+5), etc. This method was devised by Church.[84] Various systems of spiral phyllotaxis with different properties can be defined in this way. The higher the numbers of contact parastichies, e.g. (55+89), a system that one might find in a sunflower capitulum, the more nearly the angle of divergence approximates to the Fibonacci angle. This is a consequence of the very precise mathematical relationships that hold true in phyllotactic systems. The contact parastichies on a young marigold inflorescence are illustrated in Fig 6.4.

Other systems of phyllotaxis in which only one leaf occurs at each node are **distichous**, in which successive leaves are situated at an angle of 180°

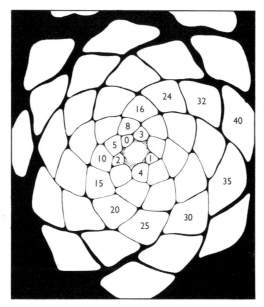

Fig. 3.1 T.S. shoot apex of a young seedling of *Pinus pinea*. Contact parastichies (5 + 8). The leaf primordia along one contact parastichy of each set are numbered, starting at primordium 0. (Adapted from Snow,[483b] Fig. 6, p. 192. After Church.[84a])

from each other, and *spirodistichous*, in which this angle is less than 180°. In *decussate* phyllotaxis, two opposite leaf primordia are formed at each node, successive pairs of primordia being at an angle of 90°; in *bijugate* systems two opposite primordia again occur at a node, but each pair is separated by an angle approximating to half the Fibonacci angle, or 68·75°. Where more than two leaf primordia are formed at a node, the phyllotaxis is said to be *whorled*. This digression on systems of phyllotaxis, some examples of which are illustrated in Fig. 3.2, is far from irrelevant, since the vascular system of the stem is composed largely of *leaf traces*, and consequently the arrangement of the leaves is directly correlated with the arrangement of vascular tissue in the stem.

Elongation of the stem is brought about by the activity of the sub-apical region, or in some instances by a true *intercalary meristem*, i.e. a meristematic region intercalated between two more differentiated regions of tissue. Such meristems are usually quite extensive regions. Increase in girth of the stem is brought about, in dicotyledons and gymnosperms, by the activity of two *lateral meristems*, the vascular cambium and the phellogen, which give rise respectively to the secondary vascular tissues and the periderm. In monocotyledons, a meristem known as the *primary*

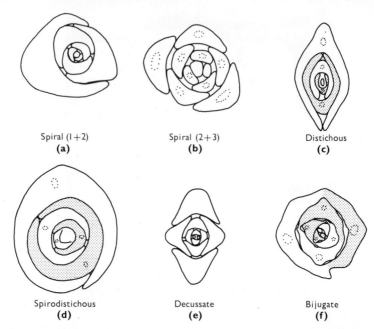

Spiral (1+2)   Spiral (2+3)   Distichous
(a)   **(b)**   **(c)**

Spirodistichous   Decussate   Bijugate
(d)   **(e)**   **(f)**

Fig. 3.2 Some examples of different types of phyllotaxis, as seen in transverse sections of the shoot apex. (**a**) *Bellis perennis*. (**b**) *Jasminum fruticans*. (**c**) *Bupleurum perfoliatum*. (**d**) *Rhoeo discolor*. (**e**) *Syringa vulgaris*. (**f**) *Dipsacus sylvestris*. In (**c**), (**d**) and (**f**) every second leaf is stippled for clarity. (After Wardlaw,[560] Fig. 43, p. 1029. (**a**), (**b**), (**c**) and (**e**) originally after Van Iterson[534a]; (**d**) originally after Snow [479a]; (**f**) originally after Snow.[483a])

thickening meristem is present in tissues below the apical meristem. Initially division of its cells contributes to the diameter of the stem, but later it participates primarily in elongation. These various meristems that bring about the growth of the stem are discussed below in more detail; secondary growth is considered in Chapter 4.

APICAL MERISTEM

The ***shoot apical meristem*** is established in the developing embryo, and may subsequently undergo considerable changes in size, shape, and rate of growth. These changes may occur throughout the ontogeny of the plant, but are most profound at the transition to the reproductive phase of development. However, throughout its development, from embryonic stages onward, the shoot apical meristem is a vital, dynamic, ever-changing, growing system; it should be mentally envisaged in its living state rather than as seen in sections, which reveal only a static arrangement of cells. Numerous experiments, involving either micro-surgical isolation of the apical meristem from adjacent tissues, or its complete excision and growth in aseptic culture, have demonstrated that it enjoys a considerable degree of autonomy with respect to the rest of the shoot, so long as it is supplied with appropriate nutrients (see Cutter [125]).

The apical meristem, which is often defined as that region of the shoot apex above (distal to) the youngest leaf primordium, is extremely variable in size and shape in different species, and often even in the same species, or the same plant, at different stages of development. It is usually radially symmetrical but in a few species it is bilaterally symmetrical. [552] The apical meristem is perhaps most commonly a paraboloid, but it may be flat or even slightly concave, variously domed, conical or even much elongated as in grasses and a number of aquatic plants. Living apices, as seen in side view, are shown in Fig. 3.3. The proximity of the leaf primordia to the tip

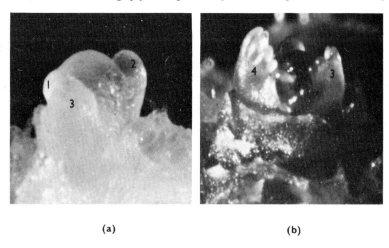

(a)   (b)

Fig. 3.3 Side view of living vegetative shoot apices. (a) Apex of the garden nasturtium, *Tropaeolum majus*. The apical dome is in the centre, with primordia 1, 2 and 3 (numbered in order of age) spirally arranged around it. Only the base of 3 remains.   × 90. (b) Apex of lupin, *Lupinus albus*. Primordia 3 and 4 already have leaflet primordia. The apical dome is again in the centre. See also Fig. 5.1a, another apex taken from a different angle.   × 75.

or centre of the apical meristem, and the relative size of the apex and the leaf primordia to which it gives rise, also vary widely (Fig. 3.4).

The width of shoot apices at the level of the youngest leaf primordium may vary in different species between about 40 μm (*Syringa*) and 3300 μm (*Cycas*). [96] In general, cycads, cacti, and some ferns have large shoot apices.

(a)   (b)

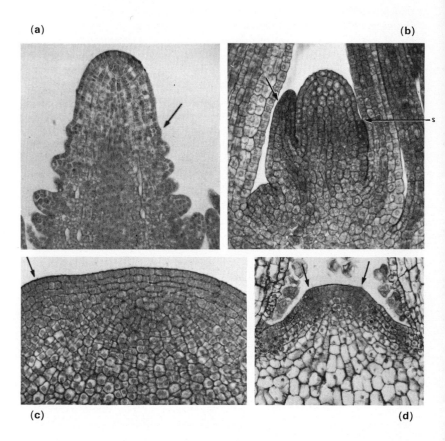

(c)   (d)

Fig. 3.4 L.S. shoot apices, showing variation in size, shape, proximity of leaf
primordia, and relative size of shoot apex and leaf primordia. The youngest leaf
primordia in the plane of section are arrowed. (**a**) *Hippuris vulgaris*, with whorled
phyllotaxis. (**b**) *Zea mays*, distichous. s, sheath of youngest leaf. (**c**) *Nuphar
lutea*, spiral. (**d**) *Perilla nankinensis*, decussate.   All × 160.

Within a single species, about a 7-fold increase in width may occur during
ontogenesis. In maize, increase in size of the apical meristem during ontogeny was found to be attributable to an increase in cell number rather
than cell size.[1]

In addition to changes that take place in the apical meristem during
ontogeny, cyclic changes occur which are related to the periodic formation

of leaf primordia. Especially in species with more than one leaf at each node, marked changes in size and form of the apical meristem may take place in the course of a single plastochrone. For example, in *Glechoma hederacea*, a species with decussate phyllotaxis, the apex increases from about 20 μm to *c*. 260 μm in height, and from *c*. 100 to *c*. 300 μm in width, during a single double plastochrone, i.e. the period required for the formation of a pair of leaves.[62] Phases of **minimal** and **maximal area**, corresponding to early and late stages of the plastochrone, are sometimes recognized. The relative proportion of its own volume to which the apical meristem gives rise during a plastochrone is related to the system of phyllotaxis,[423] emphasizing again how intimately related are the various components of the shoot. In *Trifolium*[137] and *Tradescantia*,[138] both of which have distichous phyllotaxis, the volume of the apical meristem increases more than 3-fold during a single plastochrone. In many other species, however, this increase in volume is much less noticeable.

It is thus possible to regard the shoot apex as a somewhat shifting and changing population of meristematic cells from which the growth centres of the leaf primordia become defined at regular intervals. In a recent study of the pea shoot apex,[338] the rates of cell division in different regions of the apex were obtained by measuring the increase in the number of metaphase figures after applying the drug colchicine, which arrests mitosis at that stage. Rather strangely, there was very little increase in cell division associated with leaf inception, and the rate of division was relatively constant throughout the plastochrone. However, it was found that cells are displaced from one region of the apex to another. The arbitrary regions selected are shown in Fig. 3.5c. In the first part of the plastochrone cells are displaced from the apical dome to the leaf primordium, by way of the axis (Fig. 3.5a). Later in the plastochrone the axis contributes cells to the dome (Fig. 3.5b). Thus the principal change in this apex during a plastochrone is a change in the direction of growth, as revealed by cell displacement, rather than a localized change in rates of cell division. At a particular stage during the inception of a leaf primordium there is a change in orientation of the nuclear spindles, heralding a change in the principal plane of growth.[339] The mechanisms that control these shifting directions of growth are not understood at present.

In many perennial plants, marked seasonal changes may also occur in the apical meristem. In woody plants, in particular, the shoot apex forms a terminal bud and becomes dormant (i.e. ceases growth) at certain seasons. This is an example of **episodic growth**. While many physiological changes must be involved, the main structural change lies in the development of scale leaves, or **cataphylls**, instead of foliage leaf primordia. Since the primordia are similar in early stages of development,[185] structural changes

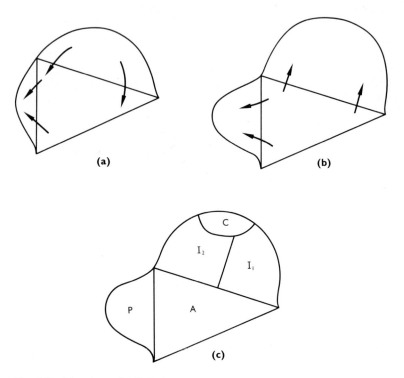

Fig. 3.5 Directions of cell displacement in the shoot apex of pea, *Pisum sativum*, during (**a**) early and (**b**) late stages of the plastochrone. (**c**) Regions into which the apex was sub-divided for analysis. A, axis; C, central zone; I_1, region where the next leaf primordium (the 10th of the plant) will be formed; I_2, region where the next primordium but one will be formed; P, leaf primordium. (From Lyndon,[338] Fig. 7, p. 14.)

in the apical meristem associated with episodic growth appear to be minimal, although both leaf formation and extension growth may cease entirely, to resume in the following spring. Both light and temperature affect bud dormancy; recent work indicates that certain hormonal substances are probably involved. Experiments indicate that bud dormancy, i.e. cessation of growth of the apical meristem and sub-apical region, may be regulated by the balance between a growth inhibitor produced in the leaves and endogenous gibberellins.[153]

In some species, dormant over-wintering buds are not formed and the whole shoot tip aborts after the onset of unfavourable conditions. Little is known about the factors controlling abortion of the tip. The first sign is usually cessation of growth;[198] eventually mitosis ceases in the shoot apex and it loses its meristematic appearance.[362] An abscission layer (see Chapter 5) forms in association with the aborting bud.

Other examples of vegetative shoot apical meristems that lose their capacity for indeterminate or unlimited growth are those meristems, often lateral buds, that develop as tendrils,[526] or thorns, e.g. *Ulex*,[43] *Bougainvillea*.[234] In *Bougainvillea* thorns result from incomplete development of the inflorescence.

In such meristems the cells lose their meristematic properties and become parenchymatous or sclerified; this is usually accompanied by considerable growth of the meristem in height. Apical meristems also become determinate structures when they are converted to a flower or inflorescence.

THEORIES OF APICAL ORGANIZATION

The shoot apical meristem is made up of a variable, but quite large, number of cells which are arranged, or organized, in various ways. Ever since the discovery of the importance of the shoot apex in 1759 by Kaspar Friedrick Wolff various *theories of apical organization* have been put forward in an attempt to describe and interpret the structure, and sometimes also the mode of growth, of the apical meristem. Most of these theories have been fully discussed in various books and reviews on the shoot apex.[75, 96, 167, 190, 203, 384] They are outlined below.

The *apical cell theory*, proposed by Nägeli in 1844, was the outcome of the discovery that there was a single tetrahedral apical cell in the shoot apex of most vascular cryptogams; it was supposed that a single initial cell of this kind would be found in the apices of all higher plants, and the theory led to extensive investigation. When it proved impossible to find an apical cell in all higher plants, this theory was superseded by the *histogen theory* proposed by Hanstein[240] in 1868, and already described, with reference to the root apex, in the preceding chapter. Hanstein considered that the shoot apex in angiosperms comprised a central core of irregularly arranged layers covered by a variable number of mantle-like layers. A single initial, or a group of initials, gave rise to each of these layers; the initials were arranged in tiers at the tip of the meristem. The *dermatogen* was the outermost layer and gave rise to the epidermis; the *periblem* was composed of the rest of the mantle layers and formed the cortex; and the inner core or *plerome* gave rise to the vascular tissue and pith (Fig. 3.6a). The weakness of this theory lies in Hanstein's attempt to assign specific destinies to the derivatives of the histogens;[190] the prospective destinies are by no means always realized. For reasons such as these, the histogen theory was rejected as a suitable interpretation of apical organization in the shoot. However,

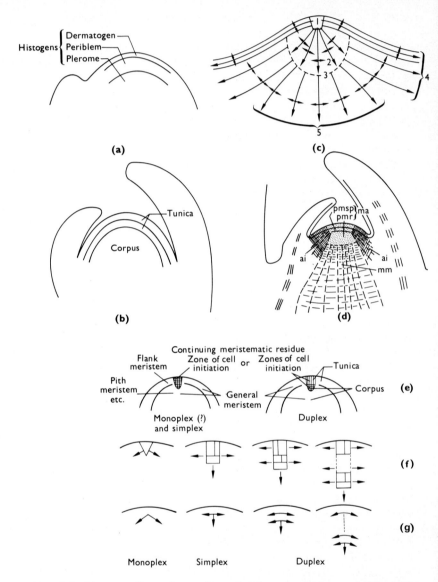

Fig. 3.6 Diagrams illustrating various theories of apical organization. (**a**) Hanstein's histogen theory, indicating the initials of the histogens. (**b**) Tunica-corpus concept. Cells in the tunica layer would divide anticlinally. (From Gifford,[203] Fig. 2, p. 480.) (**c**) Cytohistological zonation in the apex of *Ginkgo biloba*. Zone 1, apical initial group; zone 2, central mother cell zone; zone 3, represented by the broken line, indicating a transition in growth; zone 4, peripheral zone; zone 5, rib meristem. (From Gifford,[203] Fig. 1, p. 478; after Foster (1939).) (**d**) Shoot apex of

Hanstein realized the importance of the distribution of growth within the whole apex, a viewpoint which would still meet with approval at the present time.

The **tunica-corpus theory**, advanced by Schmidt[451] in 1924 on the basis of studies of angiosperm apices, was applicable only to the shoot apex and not to the root, in contrast to the two former theories. It was concerned with planes of cell division in the apex; to that extent, it may be compared with the Körper-Kappe theory of the root apex. According to this theory, the shoot apex consists of two regions or zones, the **tunica** and the **corpus** (Fig. 3.6b). The tunica consists of one or more peripheral layers of cells in which divisions are predominantly anticlinal, i.e. with walls perpendicular to the surface, except in the formation of leaf or bud primordia. The corpus consists of a central core of cells in which divisions occur in all planes. This theory is purely descriptive and carries no implication of the future destiny of the cells, though the two regions were considered to be interrelated growth zones. This theory is still often invoked in describing a shoot apex, although the two regions are variously interpreted. Some workers define the tunica very strictly and interpret other stratified layers in which occasional periclinal divisions occur (i.e. divisions in which the wall is parallel to the surface of the apex) as part of the corpus; others consider that all mantle-like layers belong to the tunica. On the latter definition the number of tunica layers in a single apex could vary from time to time. These discrepancies emphasize the need for a precise definition of terminology.

Cytohistological zonation

With the exception of the Gnetales, the apices of which have tunica and corpus regions, the shoot apices of gymnosperms do not have a surface layer which divides anticlinally and thus cannot be interpreted in terms of the tunica-corpus theory. Working with *Ginkgo biloba*, Foster[189] described the apex in terms of certain regions or zones that reacted differently to staining and were believed to have a dynamic relationship with growth processes in the apex. Modifications of this scheme were found to apply to apices of other gymnosperms, and indeed also to some angiosperms. A

Cheiranthus cheiri interpreted according to the concept of Buvat. ai, anneau initial; ma, méristème d'attente; mm, méristème médullaire; pmr, prométistème réceptaculaire; pmsp, prométistème sporogène. (From Gifford,[203] Fig. 5, p. 488; after Buvat (1952).) (e) Monoplex, simplex and duplex apices, showing how the continuing meristematic residue and general meristem can be correlated with other concepts of apical organization. (f) and (g) Diagrams showing the minimum form and minimum activity, respectively, of the basic types of continuing meristematic residue. In the right-hand diagram of the duplex apex the broken lines indicate an unspecified number of layers with anticlinal division. ((e)–(g) from Newman,[378] Fig. 5, p. 192. Published by permission of the Linnean Society of London.)

generalized diagram illustrating the various zones is shown in Fig. 3.6c. The several interrelated growth zones are considered to be derived from a group of surface initials, the *apical initial group* (1). The zone below these cells, and derived from them, constitutes the *central mother cells* (2). These cells usually stain less densely, are rather vacuolated, and are believed to divide less frequently. The centrally situated derivatives of these cells become pith, usually passing through a *rib meristem* (5) phase, and the more lateral derivatives form a densely staining *peripheral zone* (4).

Considerable variation of this generalized scheme occurs among gymnosperms and even more among angiosperms. This method of describing the shoot apex is based, not upon planes of cell division, but rather upon cell size and cytological variation as reflected in reactions to staining. The boundaries of the various zones should not be regarded as sharp.[203] Another zone which is sometimes recognized in angiosperm apices is, however, based on planes of division. This is the *cambium-like zone*, so called from a parallel alignment of cell walls which extends right across the lower part of the apical dome when seen in longitudinal section. Its presence is sometimes related to stage of the plastochrone; thus, even in the species in which it occurs, this is often a somewhat ephemeral zone.

Recent histochemical studies of apices of *Pinus* indicate that the cells in the four cytohistological zones that can be recognized in the apical meristem are not only morphologically but also biochemically different.[182] Zonation patterns of this kind are thus based on real physiological differences between localized regions of the apex. It is therefore perhaps not surprising that zonation in a particular apex may change considerably during ontogeny. That this is so has been well demonstrated in *Chenopodium*,[211] *Aster*[322] and a few other species. In the young seedling of *Chenopodium* a distinct central zone is present, characterized by lower concentrations of RNA, DNA, histone and total protein in its cells as compared with those in the peripheral zone. The regional distribution of these substances becomes much less marked as the plant ages. In pea apices, however, the amount of DNA, RNA, and protein per cell was found to be the same in all regions of the apex,[340] emphasizing the variation that may occur between species.

Anneau initial and méristème d'attente

Another interpretation of apical structure in which emphasis is placed on zones was originally based on Plantefol's theory of phyllotaxis. This theory of apical organization was put forward by Buvat[74, 75] and is supported, in particular, by many French and Belgian workers. It differs from other interpretations of apical organization and growth primarily in attributing the main histogenic role to a lateral and sub-terminal meristem, the *anneau initial* or initiating ring. The central cells of the apex, the *méristème d'attente* (waiting meristem) or **zone axiale**, are considered to be without histogenic function during the vegetative phase of development. These central cells supposedly become active in the formation of the terminal flower or inflorescence apex, where this is present. Buvat[74] postulated four main zones in the vegetative shoot apex (Fig. 3.6d). These can be correlated with the tunica and corpus as follows:

Tunica $\begin{cases} \textit{Anneau initial} \\ \text{Sporogenous promeristem} \end{cases}$

Corpus $\begin{cases} \text{Receptacular promeristem} \\ \text{Pith meristem} \end{cases}$

$\left.\begin{matrix} \\ \\ \\ \\ \end{matrix}\right\}$ *méristème d'attente*

This interpretation of apical zonation is based largely on cytological observations relating to nuclear and nucleolar size, degree of vacuolation and appearance of the chondriome, distribution of RNA as shown by specific staining, and the number and distribution of mitoses in the apical meristem. The theory is controversial chiefly because of its denial of function during vegetative growth to the centrally situated cells previously regarded as apical initials. Since the theory was first put forward its proponents have conceded that apices of all species do not conform rigidly to a single model, and the critics and supporters of the theory have moved closer in their thinking. The theory has been widely discussed[96, 123, 125, 203, 322, 555, 560] and has given rise to many experimental studies. The principal difference of opinion probably now lies not in whether the central terminal cells divide during vegetative growth but in whether they function as initials for the vegetative shoot.

Using different techniques involving direct observation of living shoot apices, various workers have shown that at least the surface cells at the tip of the apical meristem do divide, as well as those on the flanks.[29, 377] From observation of marks placed on the centre of growing apices, Loiseau[333] concluded that there was an unequal distribution of mitoses in different regions of the apex. Other workers using different techniques have also reported a lower percentage of mitoses in the distal compared to the proximal region of the apex.[41, 272, 322] In many studies of this kind it is important to have information not only on the percentage of mitoses but also on the duration and rate of mitosis, since these could differ in the various regions. Studies with tritiated thymidine followed by histoautoradiography, somewhat similar to those carried out on roots, have also demonstrated synthesis of DNA and the occurrence of divisions in the *méristème d'attente* region, though these sometimes occurred with less frequency[67, 95, 206, 324] or not at all.[493] Stimulation of mitosis can sometimes be achieved by treatment with gibberellic acid;[41] in some species this is linked with induction of flowering (see Chapter 6).

3

Various experiments have demonstrated that under certain conditions the central *méristème d'attente* region is capable of dividing actively and of giving rise to a whole vegetative shoot.[27, 123, 332] As in the case of the quiescent centre in the root, however, this does not necessarily mean that it normally does so. But on the whole the evidence for a quiescent central region in the shoot apical meristem is less strong than that for the quiescent centre in the root (see Chapter 2).

The changes that occur in the shoot apex following floral induction are discussed in Chapter 6. It has recently been suggested, however, that even in apices which remain vegetative over a considerable period of time, some changes in zonation and in the functioning and appearance of the *anneau initial* and *méristème d'attente* do occur;[386, 387] these are considered to represent an *'intermediate'* phase between the vegetative and reproductive condition (see also Chapter 6).

Types of vegetative shoot apex

In a critical discussion of apical meristems and the theories devised to interpret their functioning, Newman[378] has stated that no cells are in fact permanent initials, but that over a period of time a sequence of meristematic cells functions as initials, constituting a *'continuing meristematic residue'*, the products of successive divisions. The products of these temporary initial cells constitute the *'general meristem'*. Apices can be classified into three general types based on the form of the continuing meristematic residue (c.m.r.), according to Newman.[378] These types, illustrated in Fig. 3.6e–g, are monoplex, simplex, and duplex.

In all three types, the central meristematic residue may consist of more than one cell. In the *monoplex* apex, characteristic of ferns, the c.m.r. is in the superficial layer only. Any one cell division contributes to growth in both length and breadth; thus only one division is needed for bulk growth. In the *simplex* apex, common in gymnosperms, the c.m.r. is parallel-sided and only in the superficial layer. Both periclinal and anticlinal divisions occur; two divisions are necessary for bulk growth, but these are restricted to a single layer. In the *duplex* apex, common in angiosperms, the c.m.r. is parallel-sided and occurs in the superficial layer and at least one other successive layer. Only anticlinal divisions occur except in the innermost layer, where the cells divide in at least two planes, providing for growth both in length and breadth. Two contrasted modes of growth are thus present in the different layers, as indicated by the term duplex. This offers perhaps the simplest method yet devised of describing apices from all groups of vascular plants, whether or not they possess a tunica-corpus type of organization. An earlier classification of vegetative shoot apices into seven different types[408] has rather limited usefulness, since the variation

that exists in a single species during ontogeny, or in a single apex at different stages of the plastochrone, may transcend the differences between
types. Thus the same apex would be classified as different types at different
stages of its development. This serves to emphasize the considerable varia-
bility during ontogeny, and the difficulty of imposing any rigid classification. The simpler classification into monoplex, simplex, and duplex, based
as it is on growth processes, would probably not be subject to these diffi-
culties to the same extent.

Mutants and chimaeras

The study of the structure of mutants can be useful in elucidating the
genetic control of certain processes, and investigation of chimaeras has
thrown light on the precise origin of various tissues. Although numerous
single-gene dwarf mutants of various species are known, in general little is
known about their structure. In groundsel, however, a comparative study
showed that there were differences in size and shape between shoot apices

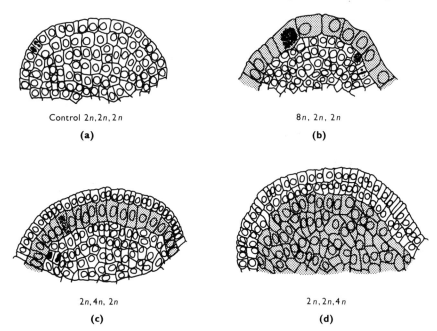

Control 2n,2n,2n   8n, 2n, 2n

(a)   (b)

2n,4n, 2n   2n,2n,4n

(c)   (d)

Fig. 3.7 Apices of axillary buds of *Datura* plants grown from seed treated with
colchicine. Various types of periclinal chimaera were obtained, of which examples
are shown. The three outermost layers of the apical meristem behave independently,
one or more becoming polyploid as a result of the treatment. Polyploid layers are
tinted. (Adapted from Satina, Blakeslee and Avery,[447] Table 1, p. 902.)

of normal and dwarf plants, these being attributable to differences in cell number.[35] Single-gene differences may thus have quite profound structural effects. Differences in ploidy, however, are less striking. Colchicine-induced tetraploid apices of *Vinca* were larger than those of diploid plants, due to a difference in cell width,[119] but otherwise little difference was noted.

In some species, *periclinal chimaeras* have been experimentally induced with the alkaloid colchicine. In such plants, the cells in different layers of the apex have different ploidy. For example, in *Datura*, apices in which the outermost three layers of cells respectively had nuclei with chromosome numbers of $8n, 2n, 2n$; $4n, 2n, 4n$; $2n, 8n, 4n$, etc. were induced (Fig. 3.7).[447] These remained stable, indicating that the two tunica layers and the uppermost layer of the corpus functioned independently of one another. It was possible to trace the tissues of lateral primordia to particular layers in the apical meristem by the ploidy level.[410] In some other genera, periclinal chimaeras are less stable.[90] Evidence concerning the functioning of the tunica and corpus can thus be obtained from these experimentally induced chimaeras.

FINE STRUCTURE OF THE APICAL MERISTEM

As yet few studies of the ultrastructure of the vegetative shoot apex have been undertaken, though some observations on apices induced to flower have been made. In general, the apical meristem consists of meristematic cells with numerous organelles; proplastids are present instead of fully developed chloroplasts.[63, 323] Even in the most highly meristematic cells some vacuoles occur, but these are small. In *Glechoma*, the cytoplasm was denser in the tunica cells than in those of the corpus, but few other differences could be observed either between these cells or between those of the central initial zone and the flank meristem.[63] In other species, a greater density of ribosomes could be observed in the cells of the flank meristem.[208] Present evidence, although as yet limited, thus reveals relatively few important differences in the fine structure of cells from the various regions of the apical meristem, even at different stages in ontogeny.

ENVIRONMENTAL EFFECTS

The structure and development of the shoot apical meristem can be affected by various environmental factors, most notably the intensity or duration of the illumination received. The most profound effects, which are brought about by daylength, lead to the induction of flowering, and are discussed in Chapter 6. Exposure to inductive photoperiods (i.e. alternating periods of darkness and light that induce the plant to flower) for a shorter time than that needed to induce flowering may result in various effects on apical growth including more rapid leaf formation. In some woody plants, stem elongation and leaf formation may be entirely arrested in short days.[565] Long day treatment may result in resumption of growth.

In some species, at least, apical size, and the organization and distribution of mitotic figures within the apex are similar whether the plants are in light or total darkness.[73, 509] In other species, increased light intensity enhances apical growth.[263] These are just a few examples of how environmental factors may profoundly affect the structure and functioning of the shoot apical meristem.

ORIGIN OF LATERAL ORGANS

Formation of leaf primordia

In most angiosperms, leaf primordia originate by periclinal divisions in the second layer of cells in the apical meristem (Fig. 5.3). This is followed by more extensive division in neighbouring tissues. In some monocotyledons periclinal divisions may take place in the surface layer of the tunica. These initial divisions occur in localized regions of the apex, according to the system of phyllotaxis; in species with decussate or whorled phyllotaxis they occur simultaneously in more than one site on the apical meristem. The leaf primordium then grows vertically upwards and finally expands laterally.

Since part of the tissue of the apical meristem is used up in the formation of a leaf primordium, the apical meristem will be in the phase of minimal area just after the inception of a leaf primordium and of maximal area just before formation of the primordium. As already pointed out, this difference in area or volume of the apical meristem is much more evident in species with decussate phyllotaxis, when two leaf primordia are formed simultaneously in opposite positions on the meristem (Fig. 3.4d).

Leaf formation and development are discussed more fully in Chapter 5.

Formation of lateral buds

In most angiosperms, lateral bud primordia occupy positions adaxial and **axillary** to leaf primordia. They are usually formed slightly later than the subtending leaf primordium, commonly when it is in its second or third plastochrone. Anticlinal divisions normally occur in the outer layers of the apical meristem; these delimit the bud meristem from the rest of the apex, and constitute the so-called **shell zone** (Fig. 3.8a, c). Periclinal divisions usually take place in the third layer of cells. Divisions in various planes ensue in the bud meristem, and it becomes organized to form an apical meristem resembling that of the parent shoot and may itself begin to give rise to leaf primordia. Commonly, however, axillary buds do not undergo much growth and development until by the ordinary processes of growth they attain a position some distance away from the parent apex.

In some species lateral bud formation does not take place until much later, when the subtending leaf is older. Such buds develop either by de-differentiation of already vacuolated cells, i.e. by the resumption of meristematic activity in differentiated cells, usually parenchyma, or from small pockets of meristematic tissue called **detached meristems**.[546] These are regions of the apical meristem which have become isolated from it by the vacuolation and differentiation of intervening tissues, and have remained dormant or inhibited until stimulated to develop into actively growing bud meristems.

Variations on this general pattern of lateral bud formation may occur. For example, axillary bud primordia may originate high up in the apical meristem, during the first plastochrone of the subtending leaf (Fig. 3.8b). However, this is unusual. Buds may not be present in the axils of all the leaves on the shoot, but may occur sporadically or in a regular sequence. In some species with opposite leaves, the two axillary buds at a node may be of unequal size and have different potentialities for further development (Fig. 3.8a).[61, 126, 466] In some species, for example gorse (*Ulex europaeus*), lateral buds may develop as **thorns**, undergoing quite profound structural changes. In *Ulex* the height of the bud apex increases rapidly by elongation of the rib meristem, and leaf and bud primordia cease to be formed. The distal cells of the shoot apex become sclerified, and various other structural changes ensue.[43] Observations such as these raise the interesting problem of what factors are involved in restricting the development of a normally indeterminate meristem, and of how they exert their effects. In recent experiments with *Ulex*, Bieniek and Millington[44] were unable to prevent development as thorns, though this could be deferred for several plastochrones. It appeared that complex hormonal regulation was involved. This problem merits further investigation.

Bud primordia may also arise adventitiously from various organs and tissues of the plant, usually by **de-differentiation** of parenchymatous cells. Most adventitious buds originate exogenously, as do axillary bud meristems, i.e. from superficial tissues. However, in some species adventitious buds may originate endogenously from deep-seated tissues.[149, 415] The problem of what causes relatively quiescent cells to divide, and become active again in this way, and especially of what causes organization of the cells into a bud meristem, is a very profound one which is still far from being understood. Hormonal factors are clearly involved, but certainly do not provide a complete answer. In the case of axillary buds, there is good evidence of correlative interaction between the bud and its subtending leaf; this relationship differs at different stages of development. In the earliest

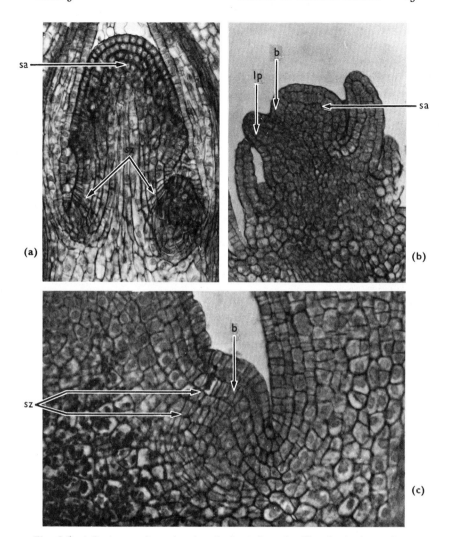

Fig. 3.8 L.S. shoot apices showing the inception of axillary buds. A two-layer tunica in the main shoot apex is evident in (a) and (b). (a) *Alternanthera philoxeroides*. Axillary bud primordia of unequal size, delimited by a clear shell zone (sz), are present in the axils of the opposite paired leaf primordia.   × 200. (b) *Hydrocharis morsus-ranae*. Inception of a bud primordium (b) in the axil of the youngest leaf primordium (lp), high up in the shoot apical meristem (sa).   × 200. (c) *Acorus calamus*. Bud in the axil of an older leaf primordium, part of which is seen on the right. Several layers of cells are involved in the inception of the bud. × 400.

stages the young leaf primordium apparently stimulates the development of its axillary bud,[481] but the mechanism involved is not understood.

STEM ELONGATION

While the shoot apical meristem gives rise to the lateral organs and is the locus of differentiation of the primary tissues, growth of the stem in length is attributable mainly to the activity of the **sub-apical region.** It has recently been suggested[435] that this region should be termed a *'primary elongating meristem'*, rather than a **rib** or **pith meristem.** Since more than the pith is involved in elongation, perhaps this new term is preferable. There is usually a rather gradual transition from the region of the apical meristem to the sub-apical region, and consequently it is perhaps misleading to consider the primary elongating meristem a true intercalary meristem.[438] Intercalary meristems are usually situated between more differentiated tissues, i.e. with a sharper boundary than is found here.

Both cell division and cell elongation contribute to growth in length of the stem, and in caulescent species both are taking place in the sub-apical region. In rosette plants there is little activity in this region and little stem elongation occurs. However, rosette plants can be caused to elongate experimentally by applying the hormone gibberellic acid (Fig. 3.9a).[437] This substance stimulates mitosis in the sub-apical region of the shoot, the induced mitotic figures being principally oriented transversely and thus leading to stem elongation. In caulescent species, such as chrysanthemum, divisions normally occur more frequently in the sub-apical region; presumably the natural content of endogenous gibberellin is higher. Treatment of such species with additional gibberellic acid has little or no effect, but mitoses can be reduced, and stem elongation much inhibited, by treatment with growth retardants.[439] Treatment with gibberellin can overcome the effects of the inhibitor on cell division and elongation (Fig. 3.9b).

These experiments thus indicate that stem elongation is normally brought about by the sub-apical region, or primary elongating meristem, and that the activity of this meristem is at least partially controlled by gibberellic acid. In some species, e.g. the gymnosperm *Ginkgo*, both **long** and **short shoots** may occur on the same plant. Morphologically, these differ primarily in the degree of internodal extension. In *Ginkgo* the apices of both long and short shoots are initially the same, but subsequently those of the developing long shoots—which also produce more auxin—possess a characteristic rib meristem, or primary elongating meristem, whereas this is absent from the short shoots.[224] There is thus a clear relationship between stem elongation and the presence of a sub-apical meristem.

True **intercalary meristems**, i.e. highly meristematic regions intercalated between two regions of differentiated, vacuolated tissue occur in the leaves and internodes of many monocotyledons, the flowering scapes

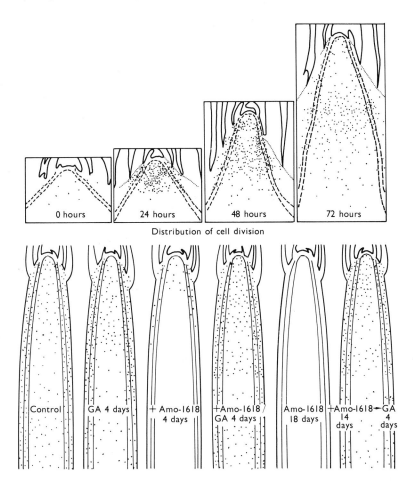

Fig. 3.9 (a) Effect of gibberellic acid (GA) on cell division in the sub-apical region of the shoot of the rosette plant *Samolus parviflorus*. 25 μg GA were applied at 0, 24, and 48 hours. Each diagram represents the median 64 μm of the shoot apex. Each mitotic figure is indicated by a dot. Boundaries of the vascular tissue are shown by broken lines. (From Sachs, Bretz and Lang,[437] Fig. 1A, p. 377.) (b) Effect of GA and the quaternary ammonium growth retardant Amo-1618 on cell division in the shoot tip of the caulescent plant *Chrysanthemum morifolium*. Transversely oriented mitotic figures (dots) in the median 60 μm of the shoot tip are shown. The various treatments, and the number of days after treatment that the material was collected or given further treatment, are indicated on the diagrams. GA can overcome the inhibitory effect of Amo-1618 on cell division. (From Sachs, Lang, Bretz and Roach,[439] Fig. 1, p. 262.)

and pedicels of some species, and the gynophore of *Arachis*, the peanut, an organ which grows downward and carries the developing fruits under the soil. Intercalary meristems of this kind may extend over a considerable area; mitotic figures are usually transversely oriented, resulting in long files of cells (Fig. 3.10). The cells are usually quite highly vacuolated and possess few of the characters normally associated with meristematic cells.

The intercalary meristem of the peanut gynophore extends over approximately 5 mm, and the region where maximal mitosis occurs corresponds with the region of maximal elongation of the organ.[269] In other dicotyledonous flowering stems, as in this one, the intercalary meristem is at the top of the stem, just below the flower. In monocotyledonous stems and leaves, however, the intercalary meristem is usually situated at or close to the base of the organ or internode. In the internodes of a number of monocotyledons, cell division and cell elongation both occur simultaneously in and above the intercalary meristem; but the later growth of the internode is restricted to cell elongation.[173, 306] In the internodes of the dicotyledons *Helianthus* and *Syringa*, where a true intercalary meristem is probably not involved, growth, involving both cell division and cell elongation, initially takes place throughout the whole internode but is later limited to successively higher levels.[578] Growth is inhibited if the leaf primordium above the internode is excised at an early stage. Work with *Helianthus*[293] indicated that young leaves were sources of the factor necessary for internode elongation. Experiments supported the view that this factor might be gibberellin or a similar substance. In the sedge *Cyperus alternifolius*, in which a single internode undergoes extensive elongation, the balance between growth by cell division in the intercalary meristem and cell elongation also changes with time.[177] Surgical experiments indicate that the stimulus leading to prolonged activity of the intercalary meristem comes from fully developed leaves, rather than axillary buds. If the erect stem or culm is decapitated, extension of the internode ceases. The stimulatory effect of the excised parts can be replaced by applied gibberellic acid and benzyladenine. Thus the prolonged activity of this meristem is apparently maintained by gibberellins and cytokinins emanating from the upper portions of the shoot; these affect both cell division and cell elongation.[178] Other evidence implicates developing leaves in the control of the plane of cell division in the stem.[375] These observations again emphasize the close relationship, both structural and physiological, between leaves and stem, the component parts of the shoot.

Also strongly indicated is the importance of the orientation of the mitotic spindle in affecting elongation of the stem, whether this is the primary factor involved or not. As yet too little is understood of the control of the plane of cell division—or, indeed, of cell elongation—and much more experimental work is needed.

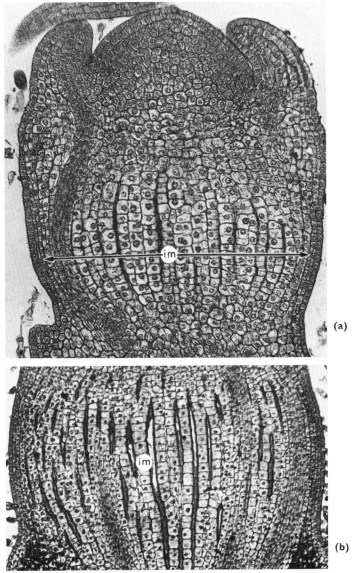

(a)

(b)

Fig. 3.10 Intercalary meristem (im) in the flower stalk of *Nuphar advena*. The stalk may ultimately elongate to several feet. (a) L.S. young developing flower primordium, showing the vertical files of cells constituting the intercalary meristem in the stalk. × 170. (b) Part of the stalk of an older flower primordium showing the intercalary meristem. × 100.

The presence of a region of meristematic, undifferentiated tissue between two mature regions of stem tissue means that vascular connections must presumably be maintained through the meristematic region. This is indeed found to be the case, the xylem usually consisting of annular and helical elements which are readily extensible.[172, 269] In most meristematic regions of the plant, of course, differentiated vascular tissues are absent.

PRIMARY TISSUES

The stem is considered to consist of three tissue systems, the *dermal*, *vascular*, and *ground tissue* components. The dermal tissue, or epidermis, differentiates from the surface layer of the shoot apical meristem, i.e. the outermost tunica layer in angiosperms. The primary vascular tissues differentiate from the procambium, which in turn develops from derivatives of the apical meristem, and the ground tissue differentiates from the peripheral or flank meristem and the rib meristem, undergoing rapid vacuolation in the process. Differentiation of vascular tissue is the most interesting and controversial of these phenomena, and is discussed in some detail below.

Differentiation of procambium

The most distal region of the apical meristem consists of cells that have dense cytoplasm and only small vacuoles, and which usually stain densely (Fig. 3.4). In a basipetal series of transverse sections of a growing vegetative shoot apex, the first evident change is the gradual vacuolation of the cells of the future ground tissue. This leaves a cylinder of more densely staining, highly cytoplasmic cells, visible as a ring in transverse section. This ring has been given various names, but is now frequently termed the *residual meristem*, from the belief that it constitutes a residuum of the meristematic tissue of the apical meristem. Within this ring more densely staining regions become evident, on proceeding basipetally in the shoot (Fig. 3.13a). These regions have a topographic relationship with the leaf primordia, and indeed constitute the procambium that develops as leaf traces. The remainder of the residual meristem eventually undergoes vacuolation and differentiates as the interfascicular parenchyma; it is thus from the derivatives of this tissue that the interfascicular vascular cambium eventually differentiates during secondary growth in many dicotyledons and gymnosperms.

The *procambium* consists of highly meristematic, densely cytoplasmic cells that are elongated in the longitudinal plane of the axis (Fig. 3.11). They undergo frequent mitoses. Studies of the fine structure of procambial cells reveal the usual features of meristematic cells, including proplastids, and a large nucleus. In most seed plants which have been studied in detail the procambium is found to be continuous with older, more differentiated vascular tissue in the stem. Careful examination of serial transverse and longitudinal sections is necessary to establish this fact,[168] since the parenchymatous leaf gaps which occur adaxial to the leaves may appear to interrupt the procambium if only a few sections are studied.

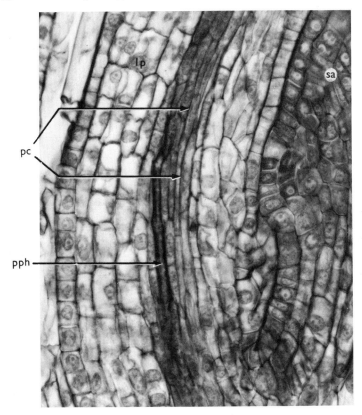

Fig. 3.11 Vascular strand in a young developing leaf primordium (lp) of *Alternanthera philoxeroides* as seen in L.S. The strand consists mainly of densely staining procambium (pc), but some protophloem (pph) has differentiated abaxially. Part of the shoot apex (sa) can be seen on the right.   × 500.

In the shoot apices of most angiosperms procambium is not found above the level of the youngest leaf primordium. In some aquatic species with an elongated apical meristem and small leaf primordia formed relatively far from the centre of the apex, and in some microphyllous cryptogams, however, procambium may be observed above the level of the youngest leaf primordia.[8] Such observations recall the classic controversy concerning whether the vascular tissue is of *foliar* or **cauline** origin. As several recent authors have pointed out,[96, 168] the distinction between cauline and foliar components of the shoot is at best somewhat artificial. Experiments in which leaf primordia were destroyed by puncturing as they became visible do, however, indicate that some procambium differentiates below an apical meristem devoid of leaves. In such treated apices, a ring of procambium uninterrupted by leaf gaps developed.[550] The occurrence of procambium, and later of differentiated vascular tissues, in the hypocotyl of the *reduced* form of the *lanceolate* mutant of tomato,[78] in which no leaves and at most only extremely rudimentary cotyledon primordia are present, also indicates that some vascular tissue can develop in the stem or its equivalent in the absence of leaves.

Some consideration may be given to the factors controlling or inducing the **differentiation** of procambium. Wardlaw[547] pointed out that procambium, or incipient vascular tissue, always differentiated in relation to an actively growing meristematic region, and suggested that some factor or factors resulting from active meristematic growth was instrumental in causing procambial differentiation. Much evidence supports this view. For example, Young[586] removed the second youngest leaf primordium from apices of *Lupinus* and found that the underlying tissue differentiated as parenchyma; if indoleacetic acid in lanolin was applied to the stump of the leaf the tissue below it remained meristematic but did not differentiate as procambium. He postulated that auxin derived from young leaf primordia caused the underlying tissue to remain meristematic, but that some other factor produced by the growing primordium, which he termed desmin, was needed to induce differentiation of procambium. Further experiments of this general nature are required.

Two somewhat opposing viewpoints—which, however, can perhaps be reconciled—have arisen because of apparently conflicting observations on the direction of differentiation of procambium. In many of the angiosperms that have been studied in detail, differentiation of procambium into the leaf primordia is **acropetal**.[165, 167] In a few angiosperms and a number of gymnosperms procambial strands have been observed in relation to the sites of leaf primordia not yet formed.[168] Recently similar observations have been reported for thorn shoots of *Ulex*.[43] The earlier observations led to the view that the procambial strand might actually determine the site of the leaf primordium.[224, 494] In experiments on both angiosperms[482] and ferns[549, 554] incisions, which would have severed any acropetally developing procambial strand or constituted a barrier between it and the primordium site, were made below incipient leaf primordia and prospective sites of future primordia. Primordia nevertheless developed in these sites, although it was known from other experiments that the positions of the youngest were not finally determined at this time;[480, 549] it was therefore concluded that the procambial strands did not induce the leaf primordia, at least in the species studied. As Allsopp[8] has pointed out, considerable metabolic activity in the apical meristem must precede actual outgrowth of the primordium; formation of a leaf primordium and differentiation of a procambial strand in relation to it may appropriately be regarded as complementary aspects of a single process.[404]

Differentiation of the procambial strands of lateral buds may be either *acropetal* or *basipetal*, according to the position of the bud. If the bud is formed close to the apical meristem, differentiation is usually acropetal; but if it develops far from the apex, from detached meristems[546] or adventitiously from parenchymatous tissues, procambial differentiation is often basipetal and indeed connection with the vascular system of the parent axis may not be established.[168] In experiments in which the central leafless region of the apical meristem of ferns[548] and angiosperms[26, 550, 551] was isolated on a plug of pith tissue, vascular connections with the parent stele being severed by the cuts, differentiation of the procambium and vascular tissue in the isolated shoot was also basipetal. These observations support the view[547] that some substance that is transported basipetally from an actively growing meristem is involved in procambial differentiation. The existence, indeed prevalence, of acropetally differentiating procambium in many species need not be considered as evidence contrary to this view. For the observations of Wetmore and Rier,[580] discussed in Part 1, Chapter 8, indicate that differentiation of vascular nodules in callus takes place at a very specific distance from a bud grafted into the callus, or from an agar source of auxin and sucrose replacing the bud. It thus appears that differentiation of vascular tissue (in this case procambium was not involved), or, by extrapolation, of procambium, may take place at a certain point on a gradient. In the case of meristems already close to differentiated procambium, this point would lie within the procambial region, and differentiation would perforce be acropetal as the source grew upwards, or appear to be so; some basipetal differentiation might add to existing procambium. But in the case of meristems at some distance from differentiated procambium— for example, detached meristems or isolated apices—differentiation would occur first at a particular point on the gradient and would then probably proceed both basipetally (especially since there would be comparatively little elongation of the axis) and acropetally. This explanation is undoubtedly over-simplified, especially since in some species some leaf traces may differentiate acropetally and others basipetally, or even bidirectionally.[168] On the other hand, it is conceivable that the point of view put forward here might indeed offer a possible interpretation of these observations.

Although dealing with fully differentiated xylem elements, rather than procambium, some experiments of Sachs[442] on decapitated pea epicotyls may be relevant here. Cuts were made in the epicotyls in such a way as partially to isolate a sliver of tissue (Fig. 3.12a). If o·1 per cent IAA in lanolin was applied to the flap of tissue, a strand of xylem differentiated in the cortical parenchyma and ran between the point of application of auxin and the vascular cylinder. If IAA was applied to the severed vascular cylinder at the same time as to the flap of tissue, however, the xylem strand failed to connect with the epicotyl vascular system (Fig. 3.12b). By means

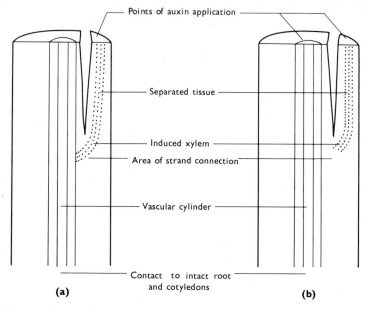

Fig. 3.12   Diagram of an experiment designed to test the influence of the concentration and direction of transport of auxin on the differentiation of xylem, using the pea epicotyl. (a) Auxin was applied only to the partially isolated flap of tissue. (b) Auxin was applied both to the flap of tissue and to the vascular cylinder. The induced strand of xylem is shown dotted. (From Sachs,[442] Fig. 2, p. 269.)

of additional experiments Sachs showed that the formation of contact between the induced and existing xylem strands depended on a difference in concentration of auxin applied to the flap of tissue and to the vascular cylinder, and the direction of the source of auxin. It seems to the writer that this result may be important in interpreting certain experiments on apical dominance, in which lateral buds released from dominance—or from a source of auxin in the main stem—both grew out and formed a vascular strand that connected with the main stele. In intact shoots, with auxin presumably being transported down the main vascular system, the bud strands often do not connect with the main stele.

Whether or not a longitudinal gradient in some substance or substances is important in controlling procambial differentiation, it seems likely that a transverse gradient across the axis may also be operative in controlling the position of the differentiating procambium or incipient vascular tissue. In apices isolated by three or four incisions on triangular or rectangular plugs of pith, the outline of the newly differentiating procambium follows the outline of the plug.[514] In a fern apex in which several radially-directed incisions had been made, a star-shaped vascular system developed at a certain level below the periphery of the tissue.[548] Thus *gradients*, perhaps of oxygen tension, from the periphery to the centre of the stem may also be of some importance in procambial differentiation. The possible importance of horizontal gradients will be discussed further in Chapter 4 in relation to the vascular cambium.

Differentiation of primary xylem and phloem

The procambium eventually differentiates into mature elements of the primary xylem and phloem, and in most dicotyledons and gymnosperms also gives rise, in part, to the vascular cambium. The first xylem and phloem elements to become mature are called the **protoxylem** and **protophloem**, respectively, later maturing elements of the primary tissue being **metaxylem** and **metaphloem**. Apart from this temporal distinction and one of size there is little difference between first-formed and later-formed elements.

In the majority of angiosperms investigated in detail, it is found that the protophloem, like the procambium itself, differentiates acropetally into the leaf primordia and is continuous with existing phloem in the axis.[165] The protoxylem, on the other hand, usually differentiates first at the base of the leaf primordium, at its junction with the axis, and then differentiates both acropetally into the leaf primordium and basipetally into the stem. The xylem is thus initially discontinuous. In shoots of *Coleus* another, slightly earlier, locus of xylem differentiation was observed in material fixed during the night; differentiation occurred first at the node below the leaf primordium, and subsequently at the base of the primordium.[271] Two initially discontinuous strands of xylem were thus differentiated. It is believed that these loci of xylem differentiation may represent sites of local increase in auxin concentration. It was also convincingly demonstrated in *Coleus* that there was a clear correlation between the amount of diffusible auxin produced by a leaf primordium and the amount of xylem differentiated in that primordium.[271] Evidence relating to the factors controlling xylem and phloem differentiation was discussed in Part 1,[127] Chapters 8 and 9.

Differentiated protophloem is usually found closer to the tip of a leaf primordium than protoxylem. Thus a basipetal series of transverse sections through a leaf trace, or sections of traces to leaves of different ages, would show only procambium in the most distal sections, then protophloem, then both protophloem and protoxylem (Fig. 3.13). Differentiation of xylem in the stem is usually *endarch*, or centrifugal, i.e. the protoxylem is nearest the centre of the stem (Fig. 3.13d); this contrasts with the

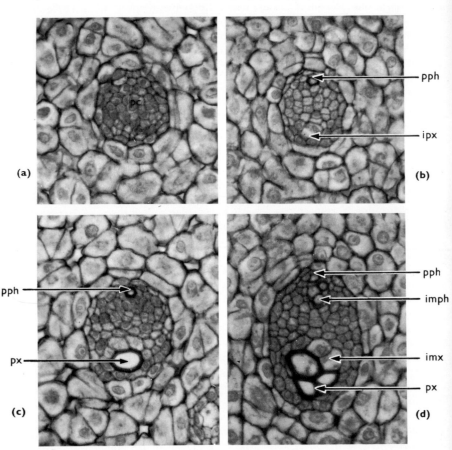

Fig. 3.13 Leaf traces in the young stem of *Acorus calamus* as seen in T.S. Figs. (a)–(d) illustrate stages in differentiation that would occur in a single vascular bundle. The centre of the stem would be towards the foot of the page. Compare procambium and protophloem as seen in L.S. in Fig. 3.11. imph, immature metaphloem; imx, immature metaxylem; ipx immature protoxylem; pc, procambium; pph, protophloem; px, protoxylem. All ×500.

exarch xylem of the root. Differentiation of phloem in the stem, as in the root, is centripetal. Thus the central region of the procambial strand is the last to differentiate, and it is from this tissue that the vascular cambium, when present, is derived.

The rate at which procambium differentiates into primary xylem and phloem differs not only in different species, but also in the same species, or even the same shoot, under different conditions. Attempts have been made to relate the plastochronal age of the first leaf primordia in which phloem differentiation and xylem differentiation first occur rather precisely to the phyllotactic system of the shoot.[165, 168, 212] It is clear that differentiation of both protophloem and protoxylem occurs later, i.e. in older leaf primordia, in shoots with higher systems of phyllotaxis; it seems likely that this is merely another interesting manifestation of the generally slower rate of leaf development characteristic of such systems. Observations such as these serve to emphasize the important relationships between structural features and rates of growth. They also demonstrate, as Wardlaw[558] has pointed out, that in high systems of phyllotaxis where many leaf primordia surrounding the apex have only procambial strands this tissue must be functionally adequate to support considerable meristematic activity. Very little is known of the manner of translocation of nutrients and other substances in tissues such as procambium or parenchyma.

In some species in which a careful analysis of the whole vascular system of the stem has been carried out, e.g. *Lupinus*,[392] it has been found that all the vascular bundles can be related to leaves or axillary buds at some level along their length; they might thus be regarded as **leaf traces**. Many leaf traces may run in the stem through several internodes, and may then connect with others. These complexes of leaf traces are sometimes called **sympodia**.

It is often observed that the leaves supplied by one sympodium of leaf traces differ in age by a number of plastochrones corresponding to a number in the Fibonacci series. From experiments which indicate that strands of xylem are inhibited from forming close to other strands that have a good source of auxin (discussed in Chapter 2, p. 32), Sachs[441] infers that vascular strands normally induced by a young leaf primordium may essentially be repelled by the strands of other leaves of similar age, and may preferentially join the strands of older leaves that have already passed the stage of maximum auxin production. This view again emphasizes the complex relationships between growth, phyllotaxis, and vascularization in the stem. Even in some stems which have continuous cylinders of vascular tissue the tissue can be shown to be initiated as discrete leaf traces. In other species, however, bundles with no connection with leaves may be present.[405] The question of the cauline or foliar origin of vascular tissue has already been discussed above with reference to procambium.

Arrangement of primary tissues

It is the variety of primary tissues and their arrangement with respect to one another that lead to the diversity of structure observed in the stems of vascular plants. This diversity is such as to make it difficult or impossible to give any adequate brief account of the structure of primary stems. Detailed systematic accounts can be found for dicotyledons in the work of Metcalfe and Chalk[360] and for monocotyledons in the books of Solereder and Meyer,[487] Metcalfe,[358] and Tomlinson.[510] A brief account of the tissues most likely to occur in mature stems of angiosperms is given here, beginning at the periphery of the stem.

The stem is bounded by the *epidermis*, a single layer of somewhat rectangular cells often covered by a *cuticle*. In stems which carry on photosynthesis *stomata* may be present in the epidermis; both covering and glandular *trichomes* may also be present. In the stems of some species a layer immediately below the epidermis differing in structure from the cortex and in origin from the epidermis may be present; this is the *hypodermis*. Below this, or directly below the epidermis, lie the cells of the *cortex*, part of the ground tissue. In many stems *collenchyma* is present near the periphery of the cortex, either forming a continuous cylinder or, more commonly, occurring discontinuously in projecting ribs. Both the collenchyma and other, parenchymatous, cortical cells, especially near the periphery of the stem, may contain abundant chloroplasts and carry on photosynthesis; such tissue is called *chlorenchyma*. Some plants, known as switch plants (e.g. *Casuarina*, *Cytisus*) have very reduced leaves and the major photosynthetic function is taken over by the stem; such

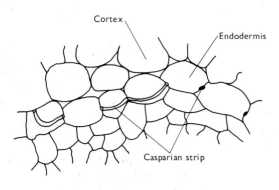

Fig. 3.14   Part of T.S. stem of *Piper*, showing Casparian strips in the endodermis.

plants may have deep furrows, bounded by chlorenchyma, in which stomata and often hairs occur. Besides collenchymatous strengthening tissue, *sclerenchyma* (usually fibres) may occur near the periphery of the stem, especially in monocotyledons.

The cells of the cortex may contain starch, or crystals, or other substances. Occasionally *idioblasts*, such as sclereids (*Trochodendron*) or oleo-resin cells (*Zingiber*), may be present. The delimitation of the cortex from the vascular tissue is usually less clear-cut than in roots, since an endodermis is usually lacking in aerial stems. An *endodermis* may occur in underground stems, or occasionally in aerial stems, such as *Piper*, where well marked Casparian strips are present (Fig. 3.14). In the young stems of certain plants, e.g. *Phaseolus*, the innermost layer of cortical cells may contain abundant starch; it is then known as a *starch sheath*. In some stems, e.g. after etiolation, the starch sheath may differentiate into an endodermis with Casparian strips.[534]

Within the vascular tissue is the rest of the ground tissue, the *pith* or medulla. These parenchymatous cells are occasionally lignified and pitted. In monocotyledons, especially, there is no real distinction between cortex and pith, and the whole is best described as *ground tissue*.

The *vascular tissue* is the source of much of the variation in stem structure. In dicotyledons the vascular tissue most commonly forms a cylinder between the cortex and the pith (Fig. 3.15a). Usually this cylinder consists of separate vascular bundles with intervening interfascicular parenchyma, but in some species the cylinder may appear to be almost continuous. In contrast to the root, with its alternating xylem and phloem, the xylem and phloem in the stem lie on the same radius, the phloem being usually external to the xylem. Such an arrangement constitutes a *collateral* vascular bundle (Fig. 3.16a). At maturity the protophloem often differentiates into thick-walled fibres.

In species of certain families, e.g. Solanaceae, Cucurbitaceae, Apocynaceae, phloem is differentiated also on the inner side of the xylem. This is known as internal phloem, to distinguish it from the external phloem in the normal position. A vascular bundle with both external and internal phloem is known as a *bicollateral* bundle (Fig. 3.16d). Other types of vascular bundle, concentric bundles, in which one type of vascular tissue encircles the other, also occur in some species. In *amphivasal* bundles, e.g. *Acorus*, *Cordyline*, the xylem encircles the phloem (Fig. 3.16b); *amphicribral* bundles, in which the phloem surrounds the xylem, are common in ferns (Fig. 3.16c). Clearly, the existence of these varied arrangements of primary xylem and phloem relative to one another renders it very difficult to envisage any simple and universally applicable mechanism controlling vascular differentiation.

In the stems of the majority of monocotyledons many collateral vascular

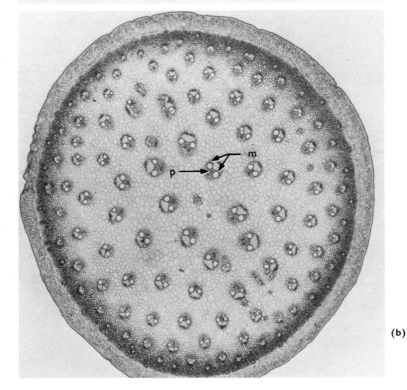

(a)

(b)

bundles are present, and these are not arranged in a cylinder but lie scattered throughout the ground tissue (Fig. 3.15b). The form of many monocotyledonous vascular bundles is such that the xylem forms the letter Y, two large metaxylem vessel elements forming the arms of the Y and the protoxylem its stem (Fig. 3.15b). The protoxylem often disintegrates, leaving a protoxylem lacuna or canal. A conspicuous bundle cap of sclerenchymatous fibres is often present, and these may encircle the bundle as well. Even the very numerous vascular bundles present in a monocotyledonous stem of this type can usually be related to leaf traces, since the typically sheathing, parallel-veined monocotyledonous leaves each have numerous leaf traces.

Anomalous structure of primary stems

Many variations of stem structure exist, both in monocotyledons and dicotyledons. Such variations are usually called anomalies, although the term may be questioned, since the so-called anomalous structure is normal for the species in question. There are monocotyledonous stems, e.g. *Coix*, in which a single ring of bundles occurs, as in a dicotyledon. There are dicotyledonous stems with scattered bundles like a monocotyledon, e.g. *Bougainvillea*. In certain dicotyledonous families rings of **cortical** or **medullary bundles** are present in addition to the normal cylinder of bundles. The medullary bundles in *Dahlia*, for example, have been shown to be leaf trace bundles which bend inwards and run down the stem in the pith for several internodes.[132] This is true also of many cortical bundles. A ring of medullary bundles is present in the stem of *Piper* and other members of the Piperaceae; in *Macropiper*, at least, the medullary bundles differentiate from a 'central cauline meristem' and have no connection with leaves except that lateral leaf traces fuse with them.[24] Other interesting patterns of vascular tissue in the stems of dicotyledons are discussed by Philipson and Balfour.[405]

Stem structure may also vary considerably depending on the habitat of the plant. In the stems of most **hydrophytes**, for example, vascular tissue is much reduced, especially xylem, and there is a well developed system of intercellular spaces. **Xerophytes**, on the other hand, have a thick cuticle, often sunken stomata, and a preponderance of sclerenchymatous tissue.

Variation in structure of stems which have undergone secondary thickening is discussed in the next chapter.

Fig. 3.15 Transverse sections of primary stems. (a) *Medicago*, dicotyledon. Vascular bundles (vb) are present in a cylinder surrounding a central pith (pi). × 50. (b) *Semele*, monocotyledon. Numerous vascular bundles are scattered through the ground tissue. m, metaxylem; p, protoxylem.   × 28.

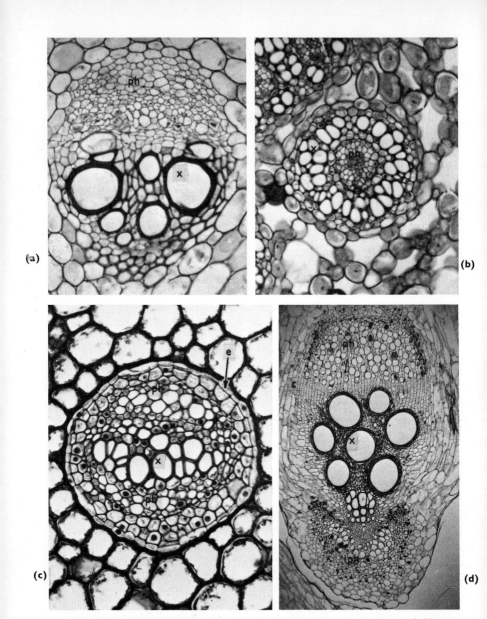

Fig. 3.16 Types of vascular bundle. (a) Collateral vascular bundle of *Piper*. × 200. (b) Amphivasal vascular bundle of *Acorus*. Xylem encircles the phloem. × 200. (c) Amphicribral vascular bundle of *Polypodium*. Phloem encircles the xylem. The vascular bundle is bounded by an endodermis. × 200. (d) Bicollateral vascular bundle of *Cucurbita*. In addition to the normal external phloem, internal phloem (iph) is present. × 25. c, cambium; e, endodermis; ph, phloem; x, xylem.

4

The Stem: Secondary Growth

The growth of the shoot is maintained not only by the apical meristem, but in many species also by the activity of lateral meristems which bring about an increase in girth. This secondary growth takes place in most dicotyledons and gymnosperms. Two lateral meristems occur: the vascular cambium, which gives rise to secondary xylem and phloem, and the phellogen or cork cambium which produces cork and phelloderm. As a result of secondary growth, the stem increases considerably in diameter and the outermost primary tissues, i.e. epidermis and sometimes also cortex, are sloughed off, the periderm forming a protective covering.

Vascular cambium

The vascular cambium usually forms a thin cylinder of cells surrounding the primary xylem. The cambium in the vascular bundles, the *fascicular cambium*, originates from the procambium remaining between the metaxylem and metaphloem. Cambium originating in the parenchymatous tissue between the bundles is known as the *interfascicular cambium* (Fig. 4.1). The cylinder of cambial cells, once complete, divides predominantly periclinally, forming a cylinder of secondary phloem towards the outside of the stem, and a cylinder of secondary xylem towards the inside. The amount of xylem formed is usually in excess of the phloem. The results of one recent experiment, in which plants of *Eucalyptus* were exposed to CO_2 labelled with ^{14}C, which was incorporated into the newly formed secondary tissues, indicate that the ratio of layers of xylem and phloem produced by the cambium was about 4:1. Environmental factors had little effect on this ratio.[543]

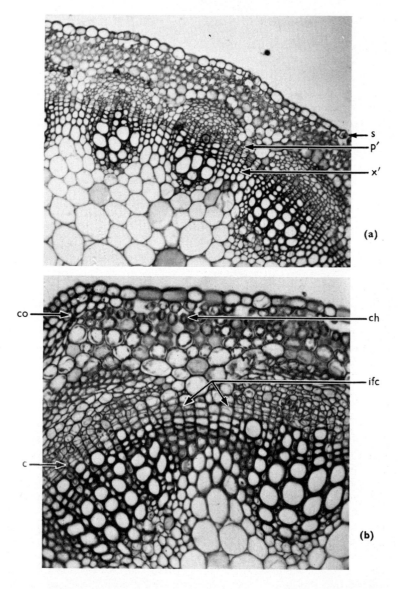

Fig. 4.1 T.S. stem of *Medicago* after secondary growth has begun. c, cambium; ch, chlorenchyma; co, collenchyma at angles of stem; ifc, interfascicular cambium; p′, secondary phloem; s, stoma; x′ secondary xylem. (**a**) ×125. (**b**) ×200.

The thin-walled cells of the vascular cambium are highly vacuolate and in this respect are unlike most other meristematic cells. Examination of cambium cells with the electron microscope confirms their highly vacuolate nature. Many ribosomes and dictyosomes, and well developed endoplasmic reticulum, are present.[492]

The cambium is made up of two kinds of cell, the *fusiform* and *ray initials*. The latter are almost isodiametric and constitute the radial system of the vascular cambium, their products differentiating as parenchymatous rays. The fusiform initials, the axial system of the cambium, are considerably elongated in the longitudinal plane of the stem and are approximately prism-shaped (Fig. 4.2). In some species the fusiform initials

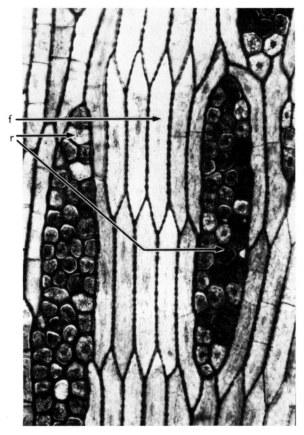

Fig. 4.2 Tangential longitudinal section through the cambial region of *Robinia pseudacacia*. f, fusiform initials; r, ray initials.   × 300.

are arranged in regular rows, having a stratified structure, in others the cambium is non-stratified. If the cambium is stratified it will give rise to stratified or *storied* wood; if it is not, the wood will be *non-storied*.

The fusiform initials of the cambium do not obey the usual laws of cell division. They normally divide vertically, in the longitudinal plane, thus contravening Errera's law, for example, which asserts that a cell will divide by a wall of minimal area. In the fusiform initial, a wall of minimal area would be transverse (horizontal). Divisions of this kind do occasionally occur, during the formation of additional ray initials, but are much less frequent than divisions resulting in a vertically oriented wall. Such divisions are mainly periclinal, but some anticlinal divisions also occur, to keep pace with the growth of the stem in girth. In longitudinal divisions the wall forms first in the region of the nucleus and grows towards the ends of the cell, which it may not reach for some time after mitosis.

By periclinal divisions of the fusiform initials, radially oriented files of cells are produced. Towards the inside of the stem these cells differentiate to form the axial system of the secondary xylem; towards the outside they differentiate into the axial system of the secondary phloem. At certain seasons a fairly wide zone of undifferentiated cells is present between the secondary xylem and phloem; these cells constitute the cambial zone, but only one layer of true cambial initials is present. Cells produced by the ray initials differentiate as the parenchymatous vascular rays. The fine structure of the fusiform and ray initials of the cambium is similar,[492] and the basis for the differences in size and shape and in the fate of their products is not yet understood.

In stems with bicollateral bundles, a cambium forms only between the xylem and the outer phloem, not between the xylem and the inner phloem.

In perennial plants cambial activity is a seasonal phenomenon and occurs during the period of active growth, beginning in the spring. In temperate climates this seasonal alternation of periods of activity and quiescence of the cambium results in the production of so-called *annual* or *growth rings* of secondary xylem and phloem. From these rings of xylem or wood the age of a tree can be calculated. Since some specimens of *Sequoiadendron*, for example, are found to be 3000 or 4000 years old, the cambial initials are evidently capable of periods of intermittent but considerable activity more or less indefinitely. This virtual immortality of the cambial cells is one of their most interesting features. Considerable research has been directed towards elucidating the factors that stimulate the seasonal activity of the vascular cambium, and some very interesting findings have recently been reported. These are discussed below, but first some consideration is given to the main product of the vascular cambium, i.e. the secondary xylem, and its varied structure.

Wood

Commercial woods consist almost entirely of secondary xylem, and are therefore a product of the vascular cambium. They are usually classified as **softwoods**, the wood of gymnosperms, principally conifers, and **hardwoods**, the wood of angiosperms, mainly dicotyledons. Softwoods consist largely of tracheids, but hardwoods contain numerous vessels.

The secondary xylem has systems of tissues that run both vertically and horizontally in the tree. For an adequate study of these systems three types of sections are required: transverse (T.S.), passing horizontally through the tree trunk; radial longitudinal (R.L.S.), passing vertically along a diameter of the tree, and tangential longitudinal (T.L.S.), passing vertically at right angles to the R.L.S. The **horizontal system**, in particular the vascular rays, appears very different in these three views, examples of which are shown in Fig. 4.3. The **vertical system** comprises tracheary elements, fibres and axial xylem parenchyma. In gymnospermous woods no vessels are present (except in the Gnetales) and the wood consists of tracheids, fibres and parenchyma, being of very uniform structure. Secretory resin ducts are often present. For further discussion of gymnosperms, see the volume by Bell and Woodcock[38] in this series. In hardwoods (except for the vessel-less Winterales) vessels are present in addition to tracheids, and the wood is much less uniform.

As explained previously, in temperate climates the seasonal activity of the vascular cambium leads to the formation of growth rings, often called annual rings, of secondary xylem. In some angiosperm genera, e.g. *Acer*, *Betula*, vessels (sometimes called pores, especially in commerce) are of approximately uniform size throughout the season and the wood is described as **diffuse porous** (Fig. 4.3d); in others, e.g. *Quercus*, *Fraxinus*, the vessels formed in the early part of the season are of much larger diameter than later-formed vessels, so that rings of wide and narrow vessels occur, and the wood is said to be **ring porous** (Fig. 4.3a). These features are most easily observed in transverse (cross) sections. In ring porous species the larger vessels are variously termed **early** or **spring wood**, and those of smaller diameter **late**, summer or **autumn wood**. The more general terms early and late wood will be used here.

Although the number of growth rings in a tree provides an approximate guide to the age of a tree, it is not always completely accurate because false growth rings may be formed within a single season as a result of a sudden check in growth, perhaps from drought or frost. Even in the same species the average width of a growth ring may vary widely according to the conditions of growth; in *Picea sitchensis* variations from 0·1 mm to about 100 times this width or more have been observed.[274]

The amount of xylem parenchyma in the axial system is very variable; frequently the cells contain starch or crystals. Timbers may be classified

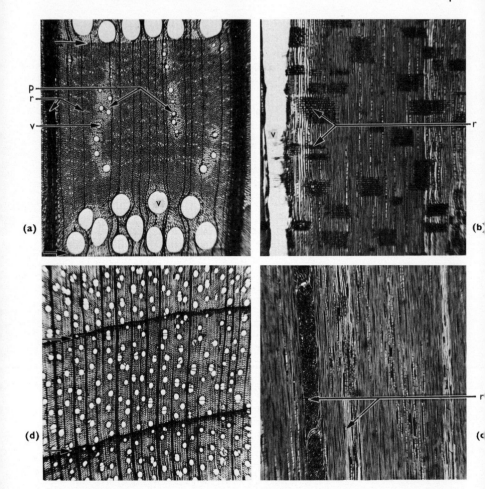

Fig. 4.3 Sections of secondary xylem. (**a**)–(**c**) T.S., R.L.S. and T.L.S. respectively of the ring porous wood of *Quercus rubra*. Vessels (v) vary greatly in size in early and late wood. (**d**) T.S. diffuse-porous wood of *Acer rubrum*. Vessels (v) are of uniform size throughout the year's growth. p, xylem parenchyma; r, rays (which vary greatly in size); v, vessel. In Figs. (**a**) and (**d**) the horizontal arrows at the left delimit a season's growth. All ×40.

according to the arrangement of axial parenchyma with respect to the vessels. *Paratracheal parenchyma* is associated topographically with the vessel elements or tracheids; *apotracheal parenchyma* is not associated with the tracheary elements. In addition, a third arrangement is

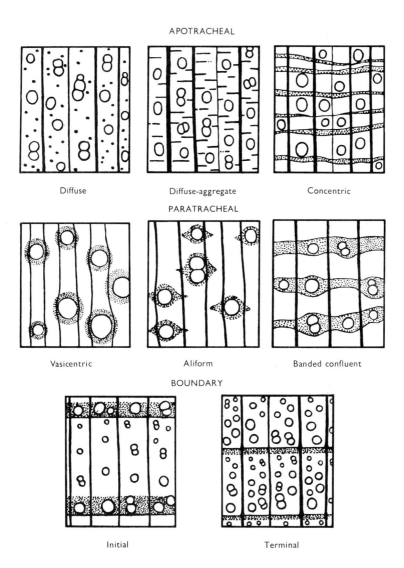

Fig. 4.4 Arrangement of axial parenchyma in secondary xylem of hardwoods, as seen in cross section. Parenchyma represented by dotted areas, except in first and second diagrams, where it is shown as isolated dots and horizontal lines respectively. (From Jane,[274] part of Fig. 58, p. 115.)

recognized, in which the parenchyma occurs primarily at the beginning or end of a growth ring and is known as **boundary parenchyma**.[274] Apotracheal parenchyma may occur as isolated strands of axial parenchyma (diffuse) or in bands arranged tangentially (metatracheal). Paratracheal parenchyma may also be sub-divided into a number of types: for example, it may encircle the tracheary elements (vasicentric), may extend laterally as wings (aliform), or may form extensive bands. Some of these arrangements are illustrated diagrammatically in Fig. 4.4.

In some timbers sapwood and heartwood can be distinguished. **Sapwood** is the still active outer secondary xylem, in which parenchymatous cells are still living. **Heartwood** is the inner xylem, often inactive and dead, and frequently darkly coloured by gums, tannins, resins, etc. Although all the cells of the heartwood are often considered to be dead, in the redwood (*Sequoia sempervirens*) it has been found that cells of the ray parenchyma may remain alive for 100 years, long after they constitute part of the heartwood.[274] The inactive tracheary elements of the heartwood often become partially or entirely blocked by **tyloses**, ingrowths of the adjacent parenchyma cells through the pits into the lumen of the vessel element. Tyloses are frequently found in the wood of *Quercus* and *Robinia* (Fig. 4.5).

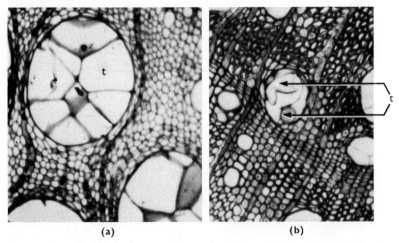

(a)   (b)

Fig. 4.5   Tyloses (t) in the wood of (a) *Quercus alba* and (b) *Robinia pseudacacia*, as seen in transverse section. (a) ×100; (b) ×200.

The horizontal system comprises the **rays**, sometimes called vascular rays, derived from the ray initials of the vascular cambium. In a block of wood cut horizontally the rays can usually be discerned as fine lines radiating out from the centre of the tree. They are seen in different perspective in T.S., R.L.S. and T.L.S. (Fig. 4.6), and all these sections are required for an understanding of their structure. Initially, at least, the rays consist of living parenchyma cells, and serve in storage and aeration. The rays may be only one cell wide (***uniseriate***) or several cells wide (***multiseriate***), a feature best observed in T.L.S. This section also reveals whether the rays are ***storied***, occurring in horizontal rows, or ***non-storied***, in a more random distribution (Fig. 4.6c). As explained previously, this is a consequence of a storied or non-storied cambium. The rays may consist of only one kind of cell (homogeneous or ***homocellular***), or of more than one kind (heterogeneous or ***heterocellular***). The cells are most frequently elongated in the radial plane, hence the name ***procumbent*** ray cells. The less common vertically elongated ray cells are known as ***upright*** cells; where present, these usually occur at the margins of the rays (Fig. 4.6).

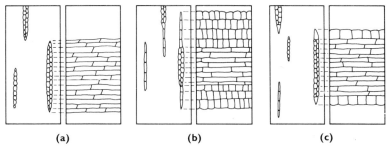

(a)　　　　　**(b)**　　　　　**(c)**

Fig. 4.6 Types of rays in hardwoods, as seen in T.L.S. (left) and R.L.S. (right). (a) Homogeneous; not exclusively uniseriate; (b) Heterogeneous; uniseriate rays composed of upright cells, multiseriate rays with uniseriate tails as long as, or longer than, the multiseriate ray. Uniseriate tails composed of upright (marginal) cells. (c) Heterogeneous, uniseriate rays of two kinds, some composed only of procumbent cells, some of upright cells. Multiseriate rays with square upright cells. (From Jane,[274] part of Fig. 62, p. 122.)

These and other features of the secondary xylem are used in the identification of woods, which is important in industry and in such fields as pharmacognosy and criminology. Minute microscopic features of wood, as of other plant structures, may prove of importance in the apprehension of criminals; wood is one of the substances most frequently left at, or accidentally taken away from, the scene of a crime.[368] The identification of timbers requires a high level of skill in plant anatomy. Wood is valued both for its variety, especially for its figure, in industry and household use, and for its uniformity; the high degree of uniformity of softwoods is one reason why they are highly valued in paper manufacture.

Factors affecting activity of the cambium

Various environmental factors affect cambial activity; many of these have a seasonal basis. For example, in *Robinia* cambial activity may be affected by daylength,[570] or by temperature.[541] Daylength affected the type of wood formed by the cambium; in short days, only late wood, typified by no or few vessels of small diameter, was formed. In long days, early wood, with more and larger vessels, was formed whether the temperature was high or low.

In conifers, the diameter of the tracheids produced by the cambium is also affected by daylength, but these effects are thought to be a secondary consequence of the effects of daylength on growth.[325] In long days needle elongation, and no doubt associated auxin production, was promoted and tracheids of large diameter (early wood) were formed. In short days tracheids of narrow diameter were formed, but if a low intensity light break was given during the dark period tracheids of wide diameter were formed, indicating that the growth response which is reflected in tracheid diameter is a true photoperiodic phenomenon.[326]

The implication of auxin as an intermediary in these phenomena was demonstrated by applying indoleacetic acid (IAA) to the cut surface of decapitated plants of *Pinus resinosa* maintained in short day conditions. In plants thus treated a zone of early wood followed the normal late wood of plants maintained in short days. Conversely, plants maintained in long days and treated with the anti-auxin tri-iodobenzoic acid (TIBA) formed a zone of tracheids of narrow diameter.[326] In *Robinia*, also, some late wood xylem was formed even in long days in seedlings treated with TIBA.[541] Other work with *Robinia*[143] indicated that early wood was formed with high levels of IAA, and late wood with low levels. These results are thus all confirmatory.

It has been known for many years that auxin could stimulate cambial activity. In 1935 Snow[483] showed that growth of the cambium was stimulated in decapitated sunflower seedlings by applied auxin, and concluded that the hormone was probably formed by young growing leaves. Over seventy-five years ago Jost[296] demonstrated that cambial growth was inhibited or prevented in disbudded shoots.

Fig. 4.7 Effects of applying various substances to disbudded shoots of *Acer pseudoplatanus* (see Appendix, p. 293). (a) Control, plain lanolin; the cambial region (c) is evident but little new tissue has been produced. (b) Treated with GA; a fairly wide, unlignified cambial zone is present. (c) IAA; some lignification of secondary xylem elements has occurred. (d) GA and IAA together; a wide zone of secondary xylem is differentiated. In (b)-(d) arrows indicate the beginning of the new growth. ((b)–(d) from Wareing, Hanney and Digby,[568] Fig. 1, p. 325. (Copyright Academic Press) ; (a) by courtesy of Professor P. F. Wareing.)

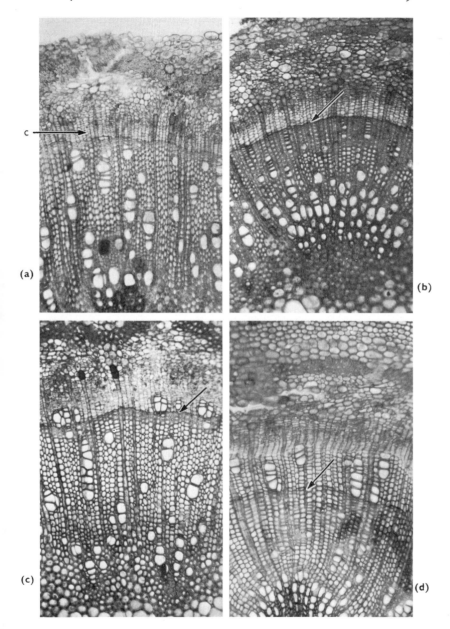

(a)

(b)

(c)

(d)

c

Interesting recent experiments, some of which can easily be carried out by a class in the laboratory (see Appendix, p. 293), have demonstrated the interaction of different hormonal substances in the control of cambial activity. If lengths of woody twigs of dicotyledons, e.g. *Populus*, are disbudded before bud break has occurred in the spring, and pure lanolin paste is applied to the apical cut surface, seasonal cambial activity is not resumed. If indoleacetic acid (IAA) is applied in the lanolin paste, some division of the cambium is induced and some differentiation and lignification of its products takes place. If gibberellic acid (GA) is applied instead of IAA, enhanced division of the cambium is induced, but few elements become lignified. If both IAA and GA are supplied, division of the cambium is stimulated and relatively normal differentiation of secondary xylem and phloem elements occurs.[566, 568] These findings are illustrated in Fig. 4.7.

A variation of this experiment, in which the material is easier for sectioning by hand, can be carried out with bean, *Phaseolus vulgaris* (see Appendix, p. 293). The growth substances are applied to a suitable internode of a decapitated plant (Fig. 4.8a), at a concentration of 10 mg/l in agar, in plastic Beem capsules (these capsules are used for embedding material for electron microscopy). The internode selected for treatment should have no cambial activity at the beginning of the experiment. The results are basically similar to those of the experiment on woody stems, i.e. application of IAA and GA together results in formation of relatively normal secondary tissues (Fig. 4.8). In this experiment the cambial region is induced to form and become active, rather than being reactivated after a period of dormancy, as in the woody twigs.

Other experiments, using both dicotyledons and gymnosperms, have confirmed the stimulatory effect of IAA on the cambium.[22] These several results suggest that in the intact shoot actively growing buds probably produce appropriate concentrations of both IAA and GA. Some recent experiments provide additional evidence of a very interesting nature.

Disbudded woody shoots of *Populus* and *Vitis* were treated, as before, with IAA and GA in lanolin paste, but the relative proportions of these substances were varied. It was found that the fate of the products of the cambium could be controlled, at least to some extent, by varying the proportions. Maximum production of xylem occurred with high IAA and low GA concentrations, whereas formation and differentiation of phloem was promoted by low IAA and high GA.[143] The balance between different growth substances thus may affect the precise site of formation and the subsequent differentiation of the cambial derivatives. These findings are supported by the demonstration[144] that in stems of both *Betula* and *Ailanthus* the level of auxin falls in short days, and that in *Ailanthus*, but not *Betula*, high levels of gibberellin remain and formation of secondary phloem apparently continues after secondary xylem production ceases.

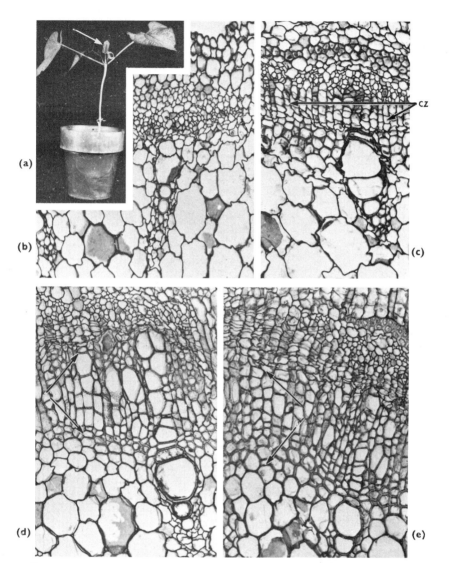

Fig. 4.8 Effects of applying various hormones in agar to decapitated plants of French bean, *Phaseolus vulgaris* (see Appendix, p. 293). (**a**) Treated plant. $\times c. \frac{3}{16}$. (**b**)–(**d**) T.S. treated internodes 7 days after treatment. (**b**) Control, plain agar. Only primary vascular tissues are present; a vascular bundle is shown. (**c**) GA. Cambial activity has been stimulated, but there is little differentiation of its products (cz). (**d**) IAA. A considerable amount of secondary xylem (arrowed) has differentiated. (**e**) GA and IAA. Considerable secondary xylem (arrowed) has differentiated, and cambial derivatives are also present on the phloem side of the cambium. (**b**)–(**e**) $\times 150$.

Subsequently extracts from the cambial zone of a ring porous and a diffuse porous species were made at various levels up the tree, and chromatograms were run and bioassayed.[144] In the diffuse porous tree, growth promoters were not present in the cambial region before swelling of the buds; at this time, when cambial activity was also resumed, a gradient of auxin was present from upper to lower levels in the tree; and later auxins were present at all levels and cambial activity also occurred throughout the whole tree. In the ring porous species, growth promoters (probably an auxin precursor) were present at all levels before bud swelling. This

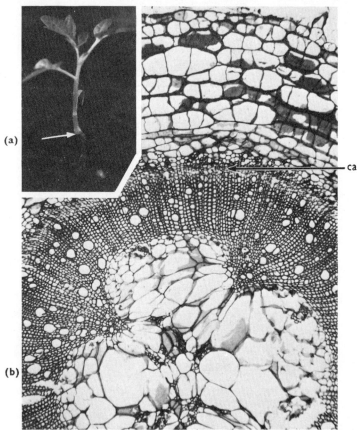

(a)

ca

(b)

Fig. 4.9   (a) A graft of normal on *reduced* tomato 18 days after grafting. Graft union arrowed. × 1. (b) T.S. hypocotyl of the *reduced* mutant of tomato which has served as scion for a normal shoot tip for 12 weeks. A vascular cambium (ca) and a considerable amount of secondary vascular tissue have been induced to develop. × 50. (From Caruso and Cutter,[79] Figs. 10 and 15, pp. 425 and 427.)

precursor is apparently converted to auxin at all levels, and cambial activity is also resumed simultaneously at all levels. A rather precise relationship between cambial activity and presence of endogenous auxin can thus be demonstrated.

In this connection, some experiments with the *reduced* mutant, a homozygous form of *lanceolate* tomato, are of interest. The *reduced* mutant grows as a green cylinder about 2 or 3 inches high, and does not produce leaves. The shoot apical meristem becomes parenchymatous at an early stage of development. The stem, or hypocotyl, of this mutant develops a primary vascular system, in which considerable cell proliferation occurs,[78] but never exhibits any cambium or secondary growth. If, however, a shoot tip of a normal, leafy tomato is grafted on to a decapitated hypocotyl of the *reduced* mutant, a vascular cambium is induced to form in the mutant and this forms abundant secondary tissue (Fig. 4.9).[79] Thus the mutant must possess the genes necessary for the formation and activity of a vascular cambium, but these are never expressed in the normal development. Presumably the formation and active growth of the cambium are induced by hormones produced by the scion of normal tomato.

It seems possible that cytokinins, usually considered to stimulate cell division, may also affect cambial activity. In cell suspension cultures derived from the cambium of *Acer*, kinetin increased cell number. The highest rate of cell division was attained in the presence of IAA, GA and kinetin together.[145] In isolated segments of etiolated pea stems, also, kinetin induced greater cambial activity and more normal secondary xylem.[488]

Some of the results of experiments on regenerating cambium also establish certain of the conditions necessary for the formation and active development of this tissue. In many split or wounded stems, cambium will regenerate from parenchymatous ground tissues and complete a continuous 'ring' (not necessarily circular) of vascular cambium (Fig. 4.10b). Theories put forward to explain these events have emphasized the importance of (1) the formation of cambium beneath and parallel to a free surface, and (2) the interruption of the original cambial ring (cylinder), the regenerating cambium tending to form a closed ring. More recently results have been obtained which would be hard to explain in terms of either of these theories, and the **gradient induction hypothesis** has been proposed.[572] It is supposed that a gradient in some factor, the nature of which is not known, tends to arise perpendicular to the exposed surface of wounded stems. In Fig. 4.10 the symbol − indicates the level of the factor nearer the exposed surface, + the level away from the surface; cambium would form at a point on the gradient lying between − and +, where the factor was at an appropriate level, and at that point only. The direction of the gradient, on this hypothesis, determines the orientation of the cambial derivatives,

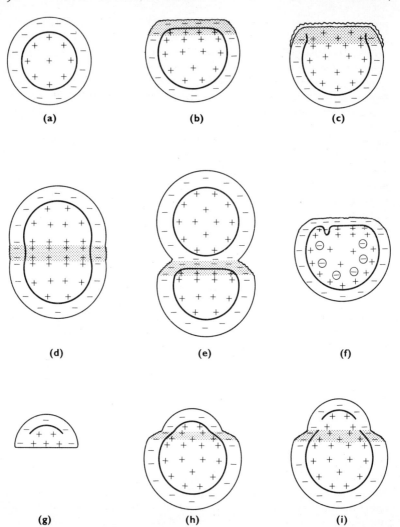

Fig. 4.10 Experiments on regeneration of cambium, interpreted in terms of the gradient induction hypothesis. See text. Cambial cells tend to form at a level inter- mediate between − and +. Regenerated tissue is stippled. (**a**) Pattern of the hypothetical gradient in a stem before operation. (**b**) Effect of excising about one- third of the stem; a new gradient from − to + is induced below the cut surface, and cambium regenerates to complete the interrupted cambial 'ring'. (**c**) If a similar wound is covered with Vaseline, the surface effect is prevented, the − level of the gradient is absent and cambial regeneration fails to take place to any extent. (**d**) Effect of splitting a stem and binding the two halves together; the surface effect

secondary phloem being formed to the − and xylem to the + side of the cambium. Once the cambium and secondary vascular tissues are differentiated, it is postulated that the level of the factor is fixed in their cells and these tend to induce similar levels in adjacent tissues. The possible importance of comparable horizontal gradients in the differentiation of procambium has been discussed in the preceding chapter.

It is possible to interpret the results of various experiments in which stems were incised and grafted together (Fig. 4.10d, e) in terms of the gradient induction hypothesis. It is particularly helpful in offering an explanation of the occasional formation of an internal cambium in wounded stems with internal phloem, e.g. Solanaceae, Campanulaceae (Fig. 4.10f).[573] Stems in which such a cambium forms in the normal development are described later in this chapter. Further experiments in which incised petioles were grafted to stems indicate that a petiole behaves like a half stem in terms of possible gradients (Fig. 4.10g, i). Cells of the ground tissue of the petiole are not competent to form cambial initials during regeneration.[574]

During normal stem development the vascular cambium is an internal tissue and may be expected to be subjected to compressive stress from adjoining tissues. The apparent importance of such pressures in controlling the normal functioning of the cambium has been demonstrated by some interesting experiments on both a gymnosperm, *Pinus strobus*, and an angiosperm, *Populus trichocarpa*. Strips of bark were released from the tree by incisions, but left attached at the apical end; they were wrapped in polythene to reduce desiccation. A pad of unorganized callus was formed on the inner side of such free strips of tissue, the ray cells being the first to proliferate. Later the callus became partially organized into radially aligned rows of cells. As in the experiments on split stems discussed above, a new cambium eventually formed, completing the cambial

---

is absent in the region of regenerated tissue, and the severed cambium merely regenerates to complete the ring. (e) Effect of grafting a cut stem and one which has been scraped or cut superficially so that the cambium is not interrupted; as a result − and + levels of the factor become juxtaposed and cambium regenerates between them, completing a cambial 'ring' in the cut stem. (f) Pattern of gradients in a Solanaceous stem which has regenerated after wounding; small circles indicate internal phloem. (g) Dorsiventral pattern of the hypothetical gradient in a petiole before operation. (h) Cut petiole/stem approach graft after regeneration to complete the cambial 'ring'. (i) Scraped petiole/stem approach graft after regeneration; cambium fails to regenerate through the ground tissue of the petiole, and a continuous ring is not formed. (From Warren Wilson, J. and P. M.,[572] Figs. 4c, e, p. 70; Warren Wilson, P. M. and J.,[573] Fig. 6a, p. 113; and Warren Wilson, P. M. and J.,[574] Fig. 6a–f, p. 13.)

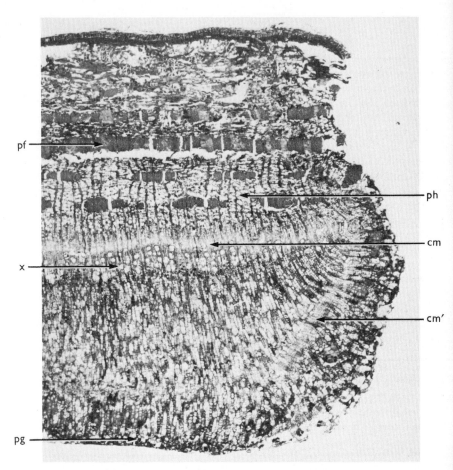

Fig. 4.11 T.S. strip of bark of *Populus trichocarpa* 21 days after the strip was freed from the tree. An extensive callus pad has developed and a new cambium (cm′) has differentiated near the tangential surface. cm, original cambium; cm′, newly differentiating cambium; pf, phloem fibres; pg, phellogen; ph, phloem; x, xylem. × 27. (From Brown and Sax,[66] Fig. 8, p. 686.)

'ring' which then gave rise normally to secondary xylem and phloem (Fig. 4.11). In similar strips of bark to which pressure was applied, by binding them to the tree with grafting bands or in other ways, the cambium functioned normally almost from the outset and a callus pad was not formed.[66] In aseptic culture, cambial explants which normally proliferated to form an unorganized callus could be prevented from proliferating by placing them

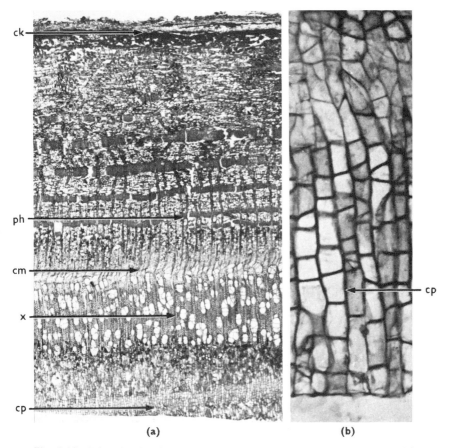

Fig. 4.12   T.S. strip of bark of *Populus trichocarpa* released from pressure for 14 days, and subsequently bound under pressure for a further 30 days. Cells in callus pad (cp) are more regularly oriented and a considerable amount of secondary xylem has been formed. ck, cork; cm, cambium; ph, phloem; x, xylem. Compare Fig. 4.11. (a) ×23, (b) ×185. (From Brown and Sax,[66] parts of Figs. 17 and 18, p. 690.)

under pressure.[65] The pressure of adjacent tissues thus seems to be a factor involved in the normal functioning of the vascular cambium (Fig. 4.12).

Reaction wood

Wood differing somewhat in structure from the normal is sometimes formed on the lower sides of branches of gymnosperms (***compression*** *wood*), and on the upper side of branches in angiosperms (***tension wood***). Both compression wood and tension wood are included in the term **reaction wood**. All of these terms have arisen because of the belief that the wood is formed as a reaction to the forces of compression or tension resulting from the position of the branch with respect to gravity. Various experiments have thrown light on the parts played by gravity and by auxin in bringing about these responses, and have contributed to our understanding of the functioning of the vascular cambium.

Branches having reaction wood have eccentric secondary xylem and phloem, indicating that an actual difference in cambial activity in different regions of the stem must be involved. Certain features are typical of reaction wood. In conifers, the tracheids of compression wood are rounded rather than angular in cross section, and intercellular spaces occur between them. They have a low cellulose content and a higher than normal amount of lignin.[113] In angiosperms, tension wood is characterized by the presence of **gelatinous fibres**. These have a thick gelatinous layer in the wall, and are usually not or only slightly lignified.[113] The gelatinous layer may sometimes undergo lignification, however.[462] There may also be a reduction in the number and size of vessels. Furthermore, the phloem in leaning stems and branches may show some anomalous features, and reaction phloem may have fibres which are less lignified but have thicker walls than usual.[562]

It is clear from these observations that in branches or trees which are at an angle to the vertical the cambium functions unevenly—more actively on the upper side in angiosperms, more actively on the lower side in gymnosperms—and its products on opposite sides of the stem also undergo rather different subsequent differentiation. Various experiments indicate that the stimulus of gravity is indeed an important factor in eliciting these responses. For example, reaction wood is not formed in plants maintained on a klinostat. The site of response to the gravitational stimulus appears to be the shoot apex.[563]

In experiments with the flexible stems of willow, it was found that if shoots were bent through 360°, forming a complete loop, and maintained in this position for a period of 11 weeks, the shoots showed greater radial growth on the upper side of both the top and bottom segments of the loops, and the gelatinous fibres typical of reaction wood were formed on this same side.[429] These results indicate that reaction wood is formed in response to gravity, and not as a result of forces of tension or compression, since these differed in the two sites where reaction wood was produced. In some ingenious experiments with conifers, the stems were tilted in opposite directions at variable time intervals. The amount of reaction wood formed in a known time could thus be calculated from sectioned stems (Fig. 4.13). It was found that under these conditions a cambial initial gave rise to approximately one cell per day. The reaction wood was recognized by

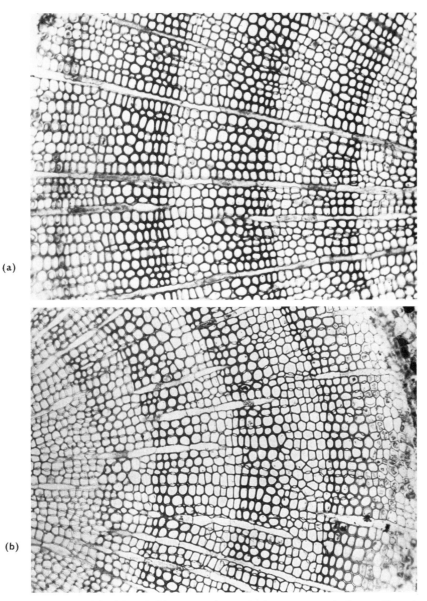

(a)

(b)

Fig. 4.13 T.S., (a) and (b) at opposite radii, of the hypocotyl of a seedling of *Larix laricina* tipped 60° to the right and left of vertical at 4-day intervals. Reaction wood is apparent as arcs of more darkly stained, heavily lignified tissue. (From Kennedy and Farrar,[309] Fig. 3, p. 423. Copyright © 1965 Syracuse University Press. Reprinted by permission of the publisher.)

excessive lignification in the tracheids, and a minimum period of 2 days was required for this to be discernible.[309]

Various experiments implicate auxin as a factor controlling the formation of reaction wood, and seem to imply that gravity may act by affecting the distribution of auxin in the stem. Experiments on gymnosperms will be considered first. Nečesaný[374] showed that if IAA in lanolin paste was applied to vertical stems of gymnosperms reaction wood was formed. Subsequent work[564] indicated that a mixture of IAA and gibberellic acid, applied to decapitated seedlings, was even more effective. Compression wood was formed symmetrically round the stem. Unilateral applications of these hormones, or of naphthaleneacetic acid (NAA) in lanolin, led to the asymmetrical development of compression wood, as well as some bending of the stems. These findings imply that reaction wood in gymnosperms is formed in response to high concentrations of auxin or similar hormones, and indeed Nečesaný[374] found that the cambial sap from the lower side of bent stems of gymnosperms, the usual site of formation of reaction wood, contained a higher amount of growth stimulators than that from the upper side.

This was true also in the angiosperms which he analysed, indicating that in this group of plants reaction wood is formed in the region with a *lower* concentration of growth stimulators and possibly a predominance of growth inhibitors, i.e. the upper side of bent stems. More recently it has been shown that in horizontally maintained stems of *Populus robusta* a greater amount of auxin (and in this case also of growth inhibitors) was present on the lower side.[328] Various experiments involving application of auxins or anti-auxins to angiosperms support the general conclusion that in angiosperms reaction wood is formed in regions with a lower concentration of auxin. Application of IAA in lanolin to the upper side of bent stems of angiosperms led to suppression of the formation of reaction wood.[374] Similarly, application of various auxins to the upper side of horizontal stems of *Ulmus*[365] and *Acer*[116] inhibited the differentiation of tension wood. Treatment of erect stems of *Acer rubrum* with auxin, however, led to the formation of a ring of tension wood below the site of application. It is believed that the concentration of auxin was sufficient to stimulate the activity of the cambium, but insufficient in inner regions of the stem to prevent the differentiation of its products as tension wood fibres.[367] The accelerated rate of cambial division may itself lead to depletion of auxin. Casperson[80, 81] has suggested that an auxin gradient may be more important than the absolute concentration of auxin.

Various experiments with anti-auxins,[115, 367] notably 2,3,5-tri-iodo-benzoic acid (TIBA), support the view that reaction wood in angiosperms is formed in regions with lower auxin concentrations. These substances are believed to block the polar transport of auxin. When TIBA was applied to

Fig. 4.14   Eight-week-old seedling of *Acer rubrum* which has been treated with a growth substance in lanolin (arrow), showing method of application to the stem. × 1·1. (From Morey and Cronshaw,[367a] Fig. 1, p. 317.)

internodes of seedlings of *Acer rubrum* in a ring of lanolin, as shown in Fig. 4.14, reaction wood, consisting of gelatinous fibres with few vessels, was formed in the stem below the treated region (Fig. 4.15).[115] A similar result was obtained with lateral application of TIBA.[366] The formation of reaction wood could be prevented by applying an auxin. In seedlings of *Ulmus americana*, also, reaction wood was formed in response to applied TIBA, cambial activity being stimulated just above the treated region, probably by an accumulation of endogenous auxin.[310, 365] The reaction wood consisted of a few wide vessels and many narrow tracheary elements and fibres (Fig. 4.16).

It is considered that the tension wood may participate actively in the reorientation of bent or horizontally maintained stems, through active

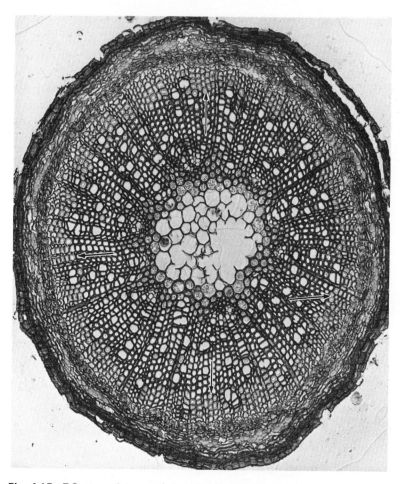

Fig. 4.15 T.S. stem of *Acer rubrum* below the region of application of 2,3,5-tri-
iodobenzoic acid (TIBA), showing the formation of a ring of tension wood
(arrows) as a result of the treatment,   ×95. (From Morey and Cronshaw,[366]
Fig. 19, p. 299.)

contraction.[116] In this connection, some interesting observations on aerial
roots of a species of *Ficus* are relevant.[589] In roots which had their tips
planted in a container of soil vigorous secondary growth, accompanied by
appreciable contraction, took place. The contraction was sufficient to lift
the container off the ground. The xylem developed during this phase
strongly resembled the tension wood formed in the stem.

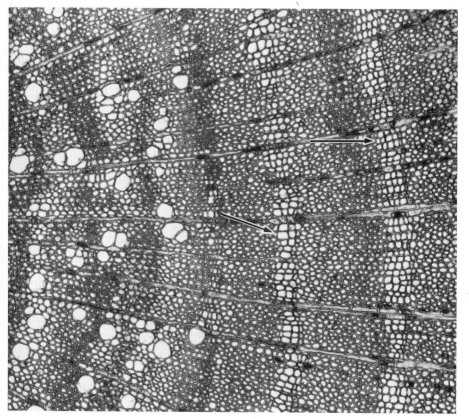

Fig. 4.16 Part of T.S. stem of *Ulmus americana* showing xylem before (left) and after treatment with 2,3,5-tri-iodobenzoic acid (TIBA). After application of TIBA the reaction xylem is characterized by tangential bands of pitted, angular tracheary elements (arrows).   × 150. (From Morey and Cronshaw,[365] Fig. 1, p. 79.)

These various observations give support to the view that the formation of reaction wood is controlled by the concentration of growth stimulating substances, notably auxin, in the cambial region, or by a change in the relative concentrations of these substances. Bending of the stem may lead to the non-uniform distribution of auxins. The notable differences in response of gymnosperms, which form reaction wood in the region of greater auxin concentration, and angiosperms, which form reaction wood in the region of lower auxin concentration, are probably a consequence of physiological differences in the cambial initials. A comparative study of cultured cambial explants from the two groups of plants might be enlightening.

Secondary growth in monocotyledons

A number of monocotyledons native to the tropics, as well as some desert plants such as *Yucca*, are arborescent, and have more or less woody stems. Some of these, notably most palms, consist entirely of primary tissues, but some show a type of secondary growth. In palms, and a number of other monocotyledons with a similar growth habit, most of the internodal tissue is a product of the ***primary thickening meristem***. This originates by periclinal divisions below the region of attachment of the young leaf primordia; it contributes to the diameter of the young stem and subsequently also to its height.[25] It is sometimes considered an end result of the juvenile phase of growth.[514] In longitudinal sections of the apical region of the shoot the primary thickening meristem can be seen to constitute several layers of rather elongated, rectangular cells oriented parallel to the surface of the stem (Fig. 4.17). The internodal regions of the young stem increase in width by the activity of this meristem.[511]

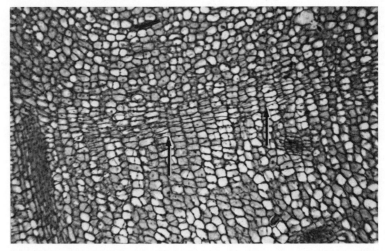

Fig. 4.17 Part of L.S. shoot apex of *Yucca*, showing the primary elongating meristem (arrowed) along the flanks of the apical region.   × 50.

In some species of monocotyledons, a type of ***vascular cambium***, apparently continuous with this meristem,[83, 159] is present in older regions of the stem and contributes to its growth in diameter. This meristem does not function like the vascular cambium in dicotyledons, but gives rise on the inner side to entire vascular bundles, comprising both xylem and

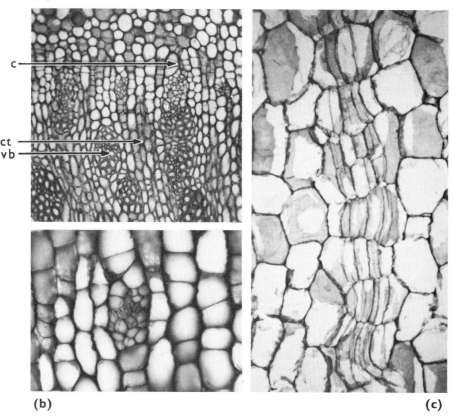

c

ct
vb

(b)   (c)

Fig. 4.18 'Cambial' region in monocotyledonous stems. (a) T.S. stem of *Dracaena*, showing the formation of entire vascular bundles (vb) embedded in pitted conjunctive tissue (ct) from the cambium (c). Outside of stem towards the top of the page. × 100. (b) Closer view of part of stem in (a) showing how a vascular bundle is formed by active division in a few cells of the cambial region. Early stages in this process can also be seen. × 300. (c) Tangential longitudinal section of the stem of *Agave* in the cambial region. Two kinds of cells are present: squat, parenchymatous cells, and narrow, elongated cells, formed by division of the former, that will give rise to a vascular bundle. × 150.

phloem in an amphivasal arrangement, embedded in a parenchymatous tissue known as ***conjunctive tissue***; to the outer side it forms only a little parenchyma.[82] The conjunctive tissue often becomes lignified. This type of cambial activity is well exemplified in stems of *Dracaena* (Fig. 4.18a) and *Cordyline*. That this type of secondary structure is adequate to support continued growth is indicated by the records of a specimen of *Dracaena draco*, which attained a girth of 45 feet and a height of over 70 feet, and was estimated to be about 6000 years old.[231] However, other observations[515] enjoin caution in estimating the age of such trees.

Very little is known about either the structure and functioning of this type of cambium, or the factors controlling its initiation and activity. The cells of the cambial region may be fusiform, rectangular or polygonal and may vary even in a single plant.[82] Fig. 4.18c shows the occurrence of two kinds of cells in the cambial region of *Agave*. The formation of a vascular bundle results from active division in various planes of the cells in a localized region of the cambium (Fig. 4.18b). In the longitudinal plane many tiers of cells are involved in the production of a single bundle.[82]

A re-investigation of secondary growth in arborescent monocotyledons is currently being conducted,[516, 588] and it is promised that later papers will demonstrate both the interdependence of primary and secondary tissues in these species and the relationship between the initiation and activity of the secondary meristem and shoot growth. It has already been observed that growing buds of *Dracaena* stimulate cambial activity in the stem below them.[515] Asymmetrical production of the wood in leaning stems of a species of *Yucca* has also been reported.[516] Moreover, in plants of *Dracaena* maintained in a horizontal position for several months there was an increase in the number of vascular bundles on the upper side of the stem, and some accompanying eccentricity of the stem, as well as increased lignification.[461] Apparently, therefore, this type of cambium responds to gravity in a manner somewhat comparable to that of a dicotyledon. The further investigation of this interesting tissue seems likely to be of great interest.

Anomalous secondary structure

Many dicotyledons, notably those with a climbing habit, show interesting secondary structure which differs from the more usual type described previously, and is therefore sometimes termed anomalous. This may be a consequence of (1) a cambium of normal (i.e. usual) type which gives rise to unusual arrangements of secondary xylem and phloem, or (2) a cambium which itself is abnormally situated and so gives rise to abnormal arrangements of tissues, or (3) the formation of accessory or additional cambia. Accounts of these stems can be found in the works of Schenck,[450] Metcalfe and Chalk[360] and Obaton.[389]

In some climbing plants, e.g. *Vitis, Clematis*, the interfascicular cambium forms only parenchyma, so that the original vascular bundles remain discrete throughout secondary growth. In *Aristolochia* and related genera this structure again occurs, with some additional features. As the vascular cylinder, broken by wide rays, increases in circumference the cylinder of sclerenchyma that encircled the bundles becomes ruptured and adjacent parenchyma grows intrusively into the gaps. Eventually a very fluted vascular cylinder is formed.

Species of *Aristolochia* are members of a biological group of woody climbers, or **lianes**, that have diverse taxonomic affinities and often show anomalous structural features, some of which may be adaptive. Among other characteristics, the vessels are often of unusually wide diameter. In some species of *Bauhinia*, a genus of woody climbers, the cambium ceases to function except in two opposite arcs, resulting in the formation of flattened, almost two-dimensional stems. More complex anomalies occur in other species.

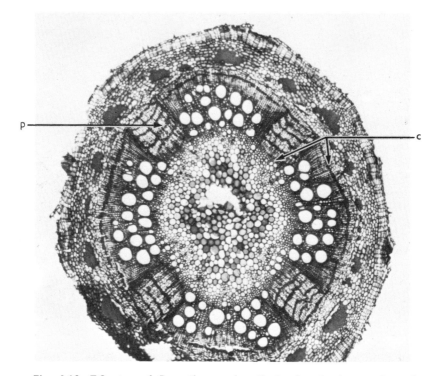

Fig. 4.19 T.S. stem of *Doxantha unguis-cati*, showing the four wedges of secondary phloem (p). In these regions the vascular cambium (c) forms more secondary phloem than secondary xylem. Section from 28th internode from the apex of a control (leafy) shoot in the experiment the results of which are shown in Fig. 4.20. ×52. (By courtesy of Dr. D. Dobbins.)

Stems of *Bignonia*, *Doxantha* and other members of the Bignoniaceae are easily recognized by the presence of wedges of secondary phloem in the cylinder of secondary xylem (Fig. 4.19). There are usually four such wedges, symmetrically arranged and corresponding in position to the larger primary

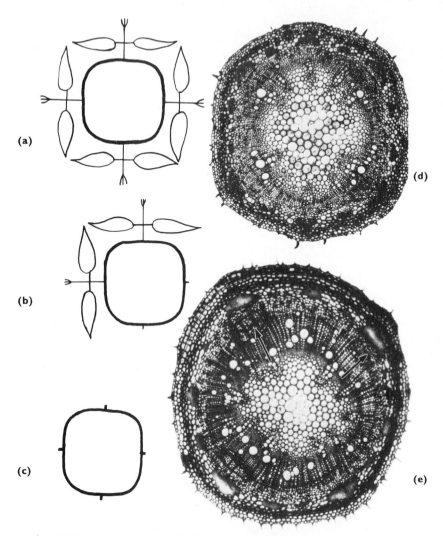

Fig. 4.20. An experiment with *Doxantha unguis-cati* in which lateral shoots were variously defoliated. (**a**)–(**c**) Diagrams of the treatments (for further details see text). (**a**) Intact shoot with 4 orthostichies of leaves. (**b**) Shoot with leaves removed from 2 orthostichies. (**c**) Shoot with all leaves removed. (**d**) T.S. 28th internode of a shoot from which all the leaves were removed. No wedges of secondary phloem developed.   × 52. (**e**) T.S. 28th internode of a shoot from which 2 rows of leaves were removed. Wedges of phloem failed to develop in association with these orthostichies, but did form in relation to the remaining leaves (arrows).   × 52. (By courtesy of Dr. D. Dobbins.)

vascular bundles.[147] During the early stages of secondary growth the vascular cambium in these four regions begins to give rise to a high proportion of phloem and correspondingly little xylem. In some species the boundary between the wedge of phloem and the adjacent secondary xylem is a straight line, in others it resembles a series of steps, the result of a sequential change in activity of small regions of cambium adjoining the original zone. In these interesting stems the vascular cambium no longer forms a continuous cylinder (Fig. 4.19), nor is its functioning uniform around the stem. The factors controlling this anomaly, and many others, require investigation. Some of the recent experimental work discussed in this chapter might suggest that at four points in the stem of *Bignonia* the concentration of gibberellin relative to that of auxin is higher, though no doubt this is an over-simplified explanation.

Experiments designed to test the effects of the leaves on the development and activity of the vascular cambium in *Doxantha* have recently been carried out by Dobbins,[148] with very interesting results. Lateral shoots resulting from removal of the main shoot tip were used as the experimental material. As the leaves emerged from the lateral bud, they were removed either from two of the four orthostichies, from all four, or all leaves were left *in situ*, as a control (Fig. 4.20 a–c). Leaf removal was continued until the shoots had produced about 30 nodes. The 28th internode of each shoot was then fixed and sectioned. In the controls the four wedges of secondary phloem had developed normally (Fig. 4.19). In shoots from which all the leaves had been removed these wedges failed to develop. Indeed, not surprisingly, little or no secondary growth occurred (Fig. 4.20d). In stems in which the leaves were removed from two orthostichies (Fig. 4.20b), more secondary growth occurred in the stem but the wedges of secondary phloem failed to occur at the sites of the two primary bundles corresponding to the treated orthostichies (Fig. 4.20e). These experiments thus demonstrate an effect of the leaves on the 'anomalous' secondary structure of the stem. In further experiments it may be possible to establish the nature of the influence of the leaves.

In other members of the Bignoniaceae, e.g. *Campsis* (Fig. 4.21), an **accessory cambium** is formed on the inner side of the normal secondary vascular cylinder.[445] This cambium forms secondary xylem elements to the outside, and secondary phloem to the inside of the stem, the inverse of the normally functioning original cambium. This may have some relevance to the gradient hypothesis of the cambium discussed earlier in this chapter. In *C. grandiflora*, at least, this cambium does not show equal activity all round the stem, but is more active at the sites of the four primary vascular bundles and in relation to the leaf positions.[238] This offers further interesting material for experimental investigation.

Climbing species of *Serjania* (Sapindaceae) have extremely interesting

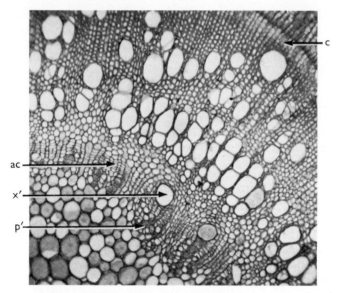

Fig. 4.21 *Campsis radicans.* The normal cambium (c), only part of which is shown, has produced secondary xylem towards the inside, and secondary phloem towards the outside, of the stem, as usual. In addition, an accessory cambium (ac) has produced secondary xylem (x′) towards the outside, and secondary phloem (p′) towards the inside of the stem, in inverse orientation.   × 100.

stem structure, the development of which has recently been studied.[286] During development the cylinder of primary vascular bundles becomes indented at certain points, so that groups of bundles become constricted off from the cylinder. This may occur even at the procambial stage. These groups of bundles behave as independent cylinders and give rise to separate cambia, each of which functions normally (Fig. 4.22). In this way the stem comes to comprise several discrete woody cylinders, each of which may ultimately even have periderm. Sometimes the woody cylinder is only lobed. It would be difficult, if not impossible, to deduce from the mature stem the manner in which this structure arose. The importance of developmental studies is thus once more emphasized. Again, the factors controlling this structure have not been investigated.

In the stems of *Bougainvillea*, which have recently been re-investigated,[170] and other members of the Nyctaginaceae, several cambia arise successively in a centrifugal direction. Each cambium produces xylem and conjunctive tissue to the inside, and phloem and conjunctive tissue to the outside. The resulting tissue gives the appearance of concentric rings of vascular bundles embedded in conjunctive tissue.

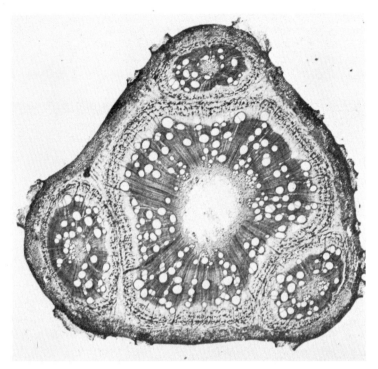

Fig. 4.22 Part of T.S. stem of *Serjania* sp. Separate vascular cylinders at the periphery have been constricted off from the central cylinder during development. Each now has an independent, normally functioning vascular cambium.   × 17.

PERIDERM

The periderm is a protective tissue which usually replaces the epidermis in stems that undergo pronounced secondary thickening. It is formed by a lateral meristem, the ***phellogen*** or cork cambium. The cells of the phellogen are meristematic, but like those of the vascular cambium are highly vacuolate; however, unlike the vascular cambium, with its fusiform and ray initials, the cells of the phellogen are all of one kind. The phellogen divides periclinally to give radially seriate files of cells; those towards the outside differentiate as ***phellem*** or cork, those towards the inside as ***phelloderm*** or secondary cortex.

In stems the phellogen most commonly originates superficially, in the sub-epidermal layer (Fig. 4.23a) or even in the epidermis itself. Periclinal divisions occur in the layer of cells concerned. The phellogen may form a

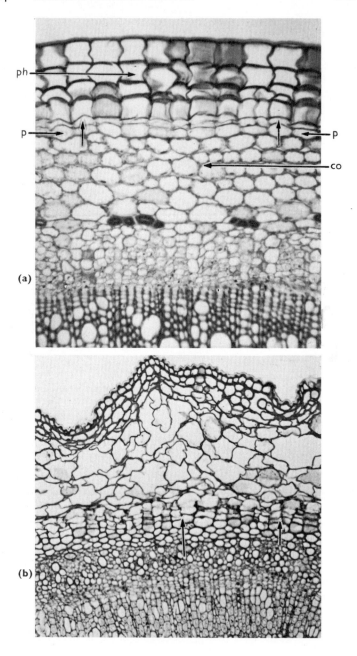

complete cylinder around the stem, or it may be formed in arcs. In some stems, e.g. *Ribes* (Fig. 4.23b), *Vitis*, the phellogen may be deep-seated, originating just outside the vascular tissue. In such instances, the cortex outside this point as well as the epidermis is eventually sloughed off and lost.

The cells of the phellem or cork are dead at maturity, and have suberized cell walls. The cells of the phelloderm, formed towards the inside of the phellogen, are living and are often distinguishable from the cortical cells only by their radial alignment with the phellogen and phellem.

The phellogen gives rise to an unequal number of derivatives on either side during a growing season, greater numbers of cells differentiating on the outer (phellem) side of the phellogen. Up to 20 rows of cork cells may be produced in a single season.

In some plants the inception and development of one phellogen may be followed by other, later-formed, successively more deep-seated phellogens. Each phellogen functions normally and produces phellem and phelloderm. Pockets of tissue, usually secondary phloem, become isolated between the periderms, and these cells die. All of this tissue is sometimes referred to as *rhytidome* (see Part 1,[127] Fig. 11.3). If the various phellogens form complete cylinders around the stem, a ring bark is formed; if they form separate arcs, a scale bark is formed.

The term *bark* includes all the tissues outside the vascular cambium, viz. secondary phloem, primary phloem, cortex, periderm and any tissues outside the periderm. Barks usually contain a quantity of sclerenchyma, including phloem fibres and often also sclereids. Indeed, barks comprise a considerable variety of tissues, and their structure is very varied in different species.[166]

Lenticels may differentiate in certain regions, usually below a stoma. In these regions the phellogen, which is continuous with the rest of the phellogen, forms a mass of loosely arranged, unsuberized cells with many intercellular spaces, the *complementary tissue*.

Cork cells are dead and because of their suberized walls are very impermeable.[109] *Quercus suber*, the cork oak, is the main source of commercial cork. The surface patterns of bark and to some extent its structure have some relationship to the growth rate of the underlying phloem.[582, 583]

In some monocotyledons *storied cork* is formed. This tissue is formed

---

Fig. 4.23 Origin of periderm. (a) T.S. stem of *Solanum dulcamara*. The phellogen (arrowed) had a sub-epidermal origin, and has given rise to one layer of phelloderm (p) and four of phellem (ph). These layers are all radially aligned. The epidermis has been sloughed off, but the presence of a wide zone of cortex (co) indicates the superficial origin of the phellogen.   × 150. (b) T.S. stem of *Ribes* sp., showing the deep-seated origin of the phellogen (arrowed).   × 200.

by periclinal divisions of superficial parenchyma cells, and does not originate from a true phellogen. The cells are radially arranged in files, and have suberized walls.

Possible factors controlling the activity of the phellogen are discussed in Part 1,[127] Chapter 11, but as yet few conclusions can be reached. A seasonal study of phellogen activity in *Robinia pseudacacia* showed that, at least in the climatic conditions prevailing in Israel, there were two periods of active growth, in April and July–August.[542] In shoots treated with GA or NAA phellogen activity occurred only in older nodes than in control plants.[19] Phellogen activity seems to be inhibited by hormones formed by an actively growing shoot apex or by growing regions of the shoot. Perhaps it may be considered as a manifestation of senescence, not unlike the abscission zone of leaves (see Chapter 5).

5

The Leaf

Leaves are lateral organs of the stem, the whole forming the shoot. As already pointed out in Chapter 3, there is a close relationship between the leaf and the stem which bears it. As Foster[188] put it, 'any effort to isolate the leaf or stem, at least from a morphogenetic standpoint, is unsupported by the facts of development'. However, the leaf does fulfil a rather specific function, for which it is highly specialized both structurally and physiologically, and it seems appropriate to devote a separate chapter to it, at the same time not forgetting that it is only one component of the leafy shoot.

The leaf is typically an organ of determinate growth, and of **dorsiventral symmetry**. Its flattened shape is well suited to its photosynthetic function, since a considerable area can be exposed to sunlight. Leaf dorsiventrality has no doubt evolved over a long period, but the ontogenetic factors controlling leaf dorsiventrality are not well understood.

Leaves may be classified morphologically as **microphylls** and **macrophylls** (megaphylls). Phylogenetically a macrophyll is a modified branch system, whereas the smaller microphyll is an enation or outgrowth of the stem which leaves no leaf gap. Ontogenetically both types of organ originate from basically similar primordia at the shoot apex, the small size of microphylls being a consequence of a failure to undergo any extensive subsequent growth.[556] Microphylls occur in the Psilotales, club mosses and some other pteridophytes.

Leaves of flowering plants vary greatly in size and form. Foliage leaves range from a few millimetres in length to over 6 feet in some palms and bananas; the floating leaves of the giant water lily, *Victoria amazonica*, may have a lamina of up to 6 feet in diameter which is capable of supporting the weight of a child or even an adult if the weight is distributed evenly over the surface of the leaf. Foliage leaves may be **simple** or **compound**; if compound, the leaflets may all originate at a central point, as in the palmate leaves of the horse-chestnut and lupin (Fig. 5.1c), or they may be formed

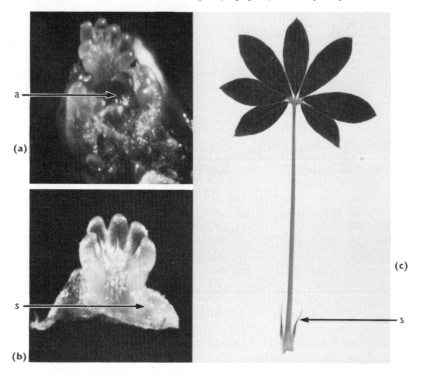

Fig. 5.1 *Lupinus albus.* (a) Shoot apex (a) surrounded by young leaf primordia P_1–P_6. The primordia of the leaflets can be seen in the older leaf primordia. × *c*. 90. (b) 6th youngest leaf, P_6, showing leaflet primordia and developing stipules (s). × *c*. 80. (c) Mature compound leaf with stipules (s). × $\frac{2}{3}$.

(usually in pairs) along a central axis or **rachis**, as in the pinnate leaves of the rose. The rose leaf is also an example of a stipulate leaf, i.e. one in which two lateral outgrowths, or **stipules**, are present at the base. In lupin small stipules also occur (Fig. 5.1c). Many foliage leaves consist of a blade or **lamina** and a stalk or **petiole**. In other species, the petiole is absent or extremely short, and the leaf is said to be **sessile**. Some leaves have a basal sheath which encircles younger leaves and the shoot apex; this is common in monocotyledons and familiar examples are the leaves of cereals and other grasses. The leaves of most dicotyledons have one main vein, or midrib, continuous with the petiole if present; lesser vascular strands form a network, and the leaf is said to have **reticulate venation**. In the leaves of most monocotyledons, several vascular bundles of more or less equivalent size form the so-called **parallel venation**. Typically, a foliage leaf subtends an axillary bud but individual leaflets of compound leaves do not have buds in their axils.

The form of the leaf often changes in a single plant during its ontogeny; such plants are said to show **heteroblastic** development.

In other species, which exhibit **homoblastic** development, leaf shape remains uniform. Leaf shape may differ in the juvenile and adult phases of the plant's development; sometimes a change in leaf shape accompanies or precedes flowering. Foliar organs which subtend flowers or inflorescences are called **bracts**, and may be modified foliage leaves or simpler, scale-like structures. Scale leaves are sometimes known as **cataphylls**, and may comprise bracts, bud scales and bulb scales. They are usually flattened, often thin structures of somewhat limited growth, with more or less sheathing bases. In the winter buds of perennial species bud scales and foliage leaf primordia are usually present, and there may be transitional forms of intermediate morphology.

In some species, photosynthetic organs which are morphologically structures other than leaves are present. For example, the entire, flattened structures present in some species of *Acacia* are not leaves but are morphologically equivalent to the flattened petiole and rachis; such structures are called **phyllodes**. In some plants, e.g. *Ruscus, Semele, Asparagus*, the photosynthetic organs actually represent flattened stems; these are called **cladodes** or **phylloclades**. They are subtended by a scale leaf.

Like the stem and root, the leaf comprises three main tissue systems: the dermal system, consisting of the upper and lower epidermis, the ground tissue, consisting of mesophyll, and the vascular system, consisting of vascular bundles forming the veins. Typically the **mesophyll**, the photosynthetic tissue, is made up of **palisade tissue**, a layer or layers of cells elongated in the transverse plane of the leaf, situated below the upper epidermis; and the **spongy mesophyll**, parenchymatous cells interspersed with abundant intercellular spaces, lying between the palisade tissue and the lower epidermis. A leaf having such a structure is said to be **dorsiventral**; one which has palisade tissue on both sides is said to be **isolateral** or isobilateral. In a few species centric leaves occur, i.e. leaves of approximately radial symmetry in which palisade tissue occurs round the whole periphery.

Several meristems, functioning either simultaneously or sequentially, contribute to the growth of the leaf. These meristems are given topographic names; they are the **apical**, **adaxial**, **marginal**, **plate** and **intercalary meristems** (Fig. 5.2). The great variety of leaf form to be found in different species, or in the same species at different times or under different

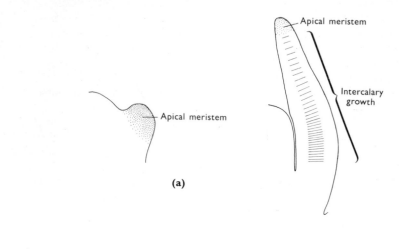

(a)

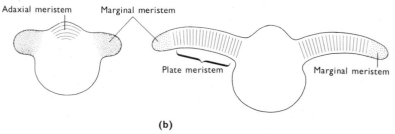

(b)

Fig. 5.2 Diagram illustrating the various meristems of a leaf primordium, as seen (a) in longitudinal and (b) in transverse section.

environmental conditions, is actually attributable to the relative activity and duration of activity of these meristems. Similarly, although this is not always made clear, experimental treatments which affect leaf shape are in fact affecting the growth of one or more of these meristems. Further research might well be directed towards interpreting experimental results in terms of meristematic activity, for this is likely to lead to a better understanding of the control of leaf development.

FORMATION, GROWTH AND DEVELOPMENT

Leaf primordia are formed in the shoot apical meristem in localized, regularly arranged sites according to the phyllotaxis, as explained in Chapter 3. These sites are on the flanks of the apical meristem, at a variable distance from the tip (see Chapter 3). During a *plastochrone*, i.e. the period between the formation of one leaf primordium and the next, the shoot apex fluctuates in size, the amount depending on the size of the leaf primordium relative to that of the shoot apex. Various events occur in the apical meristem prior to the appearance of a leaf primordium; however, despite the generality of the phenomenon of leaf formation in the plant kingdom, these events are not, as yet, well understood. Some examples of the complexities which may occur are cited. In the shoot apex of *Narcissus*, for example, the number of cells in division on the flank where the next leaf primordium will be formed is approximately double that on the opposite flank.[135] In *Pisum*, the pea, displacement of cells and changes in the plane of growth occur in the apical dome 16 hours before the beginning of the plastochrone, i.e. before the time that the primordium appears.[338, 339] Histochemical studies have shown an increase in RNA at the site of a young leaf primordium,[211] and other biochemical changes must also occur.

The first structural sign of leaf inception is a periclinal division on the apical flank, usually in the second or third layer of cells (Figs. 5.3, 5.4c). In monocotyledons, periclinal divisions may occur in the superficial tunica layer; in dicotyledons, the outermost tunica layer usually divides only anticlinally and gives rise to the epidermis of the young leaf primordium. The initial periclinal division is usually followed by further division of adjacent and underlying cells (Fig. 5.3), resulting eventually in elevation of the leaf primordium above the surface of the apical meristem.

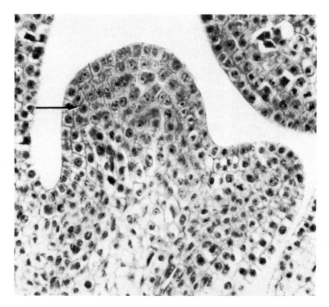

Fig. 5.3 Leaf inception in pea (*Pisum sativum*). The arrow indicates a region of leaf inception; a periclinal division can be seen in the second layer.   × 300.

5

In species with sheathing leaf bases, such as grasses, divisions spread laterally from the original localized site on the apical meristem, resulting in a crescent-shaped primordium which is not yet of dorsiventral symmetry but is thickest at the point of origin and tapers to either side.[305] The primordium begins to grow vertically, and divisions also spread further around the apex, leading to the upgrowth of a collar or rim of tissue round the circumference of the shoot apex (Fig. 5.4 a and b). This is shorter than the original locus of growth, the apex of the leaf primordium. One consequence of this encircling growth is that median longitudinal sections of a

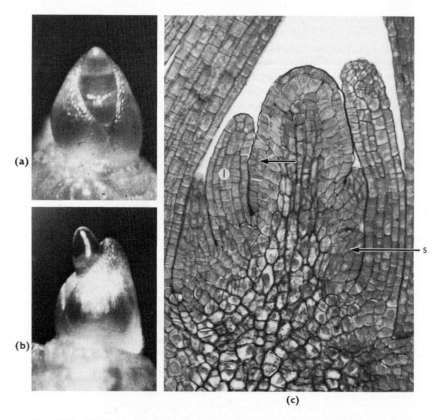

(a)

(b)

(c)

Fig. 5.4 Shoot apices and developing leaf primordia of *Triticum aestivum*. (a) and (b) Living shoot apices viewed from different angles, showing the ensheathing leaf primordia. Older primordia have been removed, so that the tip of the shoot apex now protrudes.   × 50. (c) L.S. shoot apex. The arrow indicates periclinal divisions which will result in the formation of a new leaf primordium. l, leaf primordium; s, sheath.   × 200. (By courtesy of Dr. B. C. Sharman.)

leaf primordium on the apex also pass through the sheath on the opposite side of the apex (Fig. 5.4c).

Axillary bud primordia are not usually formed until the leaf primordium is in its second or third plastochrone (see Chapter 3). In species with stipulate leaves, small outgrowths are formed on either side of the primordium base usually in the second or third plastochrone (Fig. 5.1a, b). In species with palmately compound leaves, such as lupin (Fig. 5.1), the leaflet primordia also become evident in about the third plastochrone. They are formed as a result of the establishment of several centres of growth with intervening areas where there is little growth. In pinnate leaves the pinnae develop from regions of the marginal meristem, and their inception is usually delayed until a later plastochrone.

Apical and intercalary growth

In the early stages after its inception, for example, in rice during the second plastochrone,[305] the primordium manifests *apical growth*. It thus forms an elongating structure projecting from the apical meristem (Fig. 5.3). In peas, the increase in length of a primordium during a plastochrone is estimated at 60 μm.[337] In many species, vertical growth is greater on the abaxial side of the primordium, i.e. on the side away from the axis or shoot apex. In consequence, young primordia often bend over the apical meristem, conferring upon it some degree of protection.

In ferns, apical growth of the leaf is prolonged and there is a distinctive apical cell. In the leaves of angiosperms, there are usually no definite apical initials and apical growth ceases relatively early in the development of the young primordium.[188] Subsequent extension of the leaf is a result of general cell division and enlargement throughout the primordium. Cell division ceases first at the tip of the leaf, and last at the base.[346] Extension of the leaf is thus brought about by intercalary growth (Fig. 5.5), except in the early stages. In most monocotyledons there is a fairly well defined *intercalary meristem* at the base of the leaf primordium. Such a meristem is present also in the leaves of water lilies (Fig. 5.6), and is responsible for the elongation of the petiole which raises the lamina above the water surface. Division proceeds in such a way as to form longitudinal files of cells contributing to the extension of the leaf (Fig. 5.6b). In foliage leaves of *Narcissus* apparently there is no apical growth in the primordium, and all elongation is due to intercalary activity.[136]

Marginal growth

When the young dorsiventral leaf primordium attains a certain height small bulges or outgrowths appear laterally on either side (Fig. 5.7a). These constitute the *marginal meristems*, which by their activity

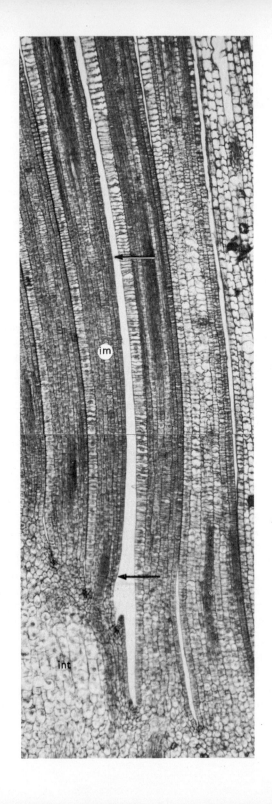

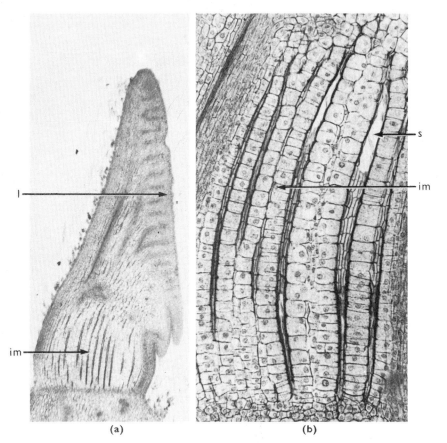

Fig. 5.6 L.S. developing leaf primordium of *Nuphar lutea*. (a) L.S. whole primordium. im, intercalary meristem; l, rolled-up leaf lamina.   × 40. (b) L.S. part of intercalary meristem (im), with air spaces (s). Small-celled procambial tissue is also present.   × 150.

Fig. 5.5 L.S. base of developing leaf primordia of *Cyperus alternifolius*. The approximate extent of the intercalary meristem (im) in one leaf is indicated by arrows. The shoot apex is to the left of the picture. int, developing internode. × 150. (Slide by courtesy of Dr. J. B. Fisher.)

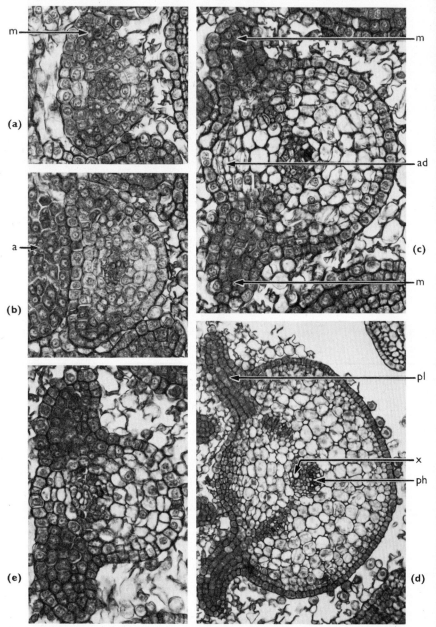

Fig. 5.7 T.S. developing leaf primordia of alligator weed, *Alternanthera philoxeroides*. The shoot apex is to the left of each photograph. (a)–(d) Leaf primordia from a single shoot apex. (a) The 2nd youngest primordium, P₂. (b) The

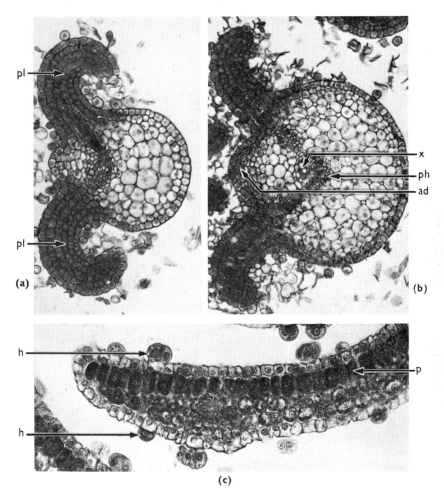

Fig. 5.8 T.S. developing leaves. (a) and (b) Sections at different levels of a P₃ of *Alternanthera*, (b) being more proximal.   × 150. (c) T.S. leaf primordium of *Callitriche* sp. The floating leaf lamina of this aquatic plant is thin and shows early differentiation of palisade tissue (p), as well as numerous glandular hairs (h). × 300. ad, adaxial meristem; ph, phloem; pl, plate meristem; x, xylem.

same primordium more proximally. (c) The 3rd youngest primordium, P₃, with a well-developed adaxial meristem. (d) P₄, from the same apex. Protophloem and protoxylem are differentiated in the midrib bundle, and a plate meristem (pl) is being formed. (e) A late P₂ from a different apex. a, shoot apex; ad, adaxial meristem; m, marginal meristem; ph, protophloem; pl, plate meristem; x, protoxylem. (a)–(c) and (e), × 300; (d), × 150.

establish the number of layers of mesophyll cells in the lamina. The marginal meristem first becomes distinguishable at different stages according to the species. In *Xanthium* the leaf primordium is about o·22 mm at the initiation of marginal growth;[348] the marginal meristem maintains its activity for a period of about 23 days,[349] and 6 layers are established in the leaf.[345] In

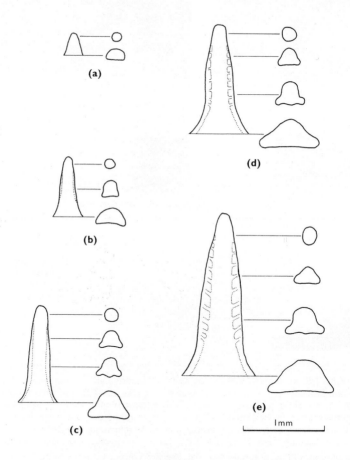

Fig. 5.9 Diagram of T.S. and L.S. leaf primordia of tobacco, in successive stages of development. Development of the lamina is indicated by dotted lines; in (**d**) and (**e**) the positions of the main lateral veins are also thus indicated. (From Avery,[20] Figs. 1 and 3–6, p. 566.)

transverse sections of young leaf primordia the marginal meristems are evident at the edges of the flanges of tissue projecting from the future midrib (Fig. 5.7 and 5.8). In Fig. 5.9 and 5.10 the development of the leaf

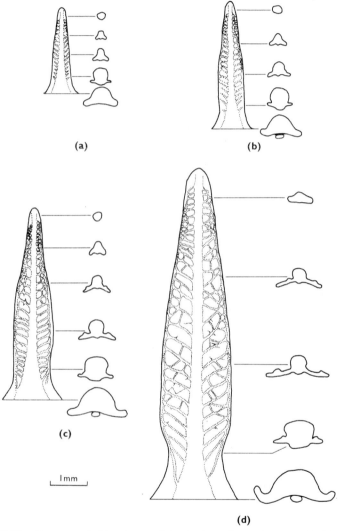

Fig. 5.10 Diagrams of older stages of developing tobacco leaves, as in Fig. 5.9. In (a) the primordium is approximately 2·25 mm long, with the procambium blocked out at the distal end. In (c) the primordium is about 5 mm long. The system of venation is seen in (d). (From Avery,[20] Figs. 18–21, p. 572.)

margins and vasculature in tobacco leaves is indicated diagrammatically, as it would be seen in longitudinal view and in transverse section at various levels.

In some species distinctive marginal and sub-marginal initials have been recognized; in such instances, diagrams which project the future destiny of the derivatives of these cells can be prepared.[188] In other species, however, marginal growth is much more variable and cannot be interpreted by a simple diagram;[414] indeed, it may not be possible to distinguish marginal and sub-marginal initials.[139] Hara[241] has recognized two major types of marginal growth in dicotyledonous leaves—marginal and sub-marginal—and has sub-divided the latter into three types, adaxial, abaxial and middle, according to which layer gives rise to the procambium. However, he reports that the leaves of a particular species do not always show the same type of marginal growth; this may even vary in a single lamina. This situation is reminiscent of that concerning the classification of shoot apical meristems discussed in Chapter 3; probably neither type of organized growth is sufficiently stable for a classification to be very meaningful.

In leaves of grasses, marginal and apical growth are not distinct processes;[305] the leaves of *Narcissus* also have no distinct marginal growth,[136] but in broad-leaved monocotyledons marginal growth does occur.[414] This might indeed be expected from the form of the mature leaf, for example in genera such as *Hosta* and the water hyacinth, *Eichhornia*.

Adaxial growth

In leaves of many dicotyledons a strip of cells below the adaxial epidermis divides periclinally and contributes to the thickness of the leaf.[188] This **adaxial meristem** (Fig. 5.2b, 5.7c) is usually centrally situated and contributes especially to growth in thickness of the petiole and midrib. In some foliar organs, for example the phyllode of *Acacia*[47] and the leaf of *Acorus*, discussed in more detail below, this adaxial meristem is extremely active and results in the formation of a lamina-like structure which is laterally, instead of dorsiventrally, flattened (Figs. 5.11 and 5.12).

A recent study examines this adaxial growth in more detail, and also throws some light on the genetic control of leaf growth. The so-called unifacial leaves of *Acorus* and some other monocotyledons are flattened laterally, i.e. the blade is extended in the radial plane (Fig. 5.12). In *Acorus* the leaf apex matures into an attenuated tip at an early stage, and subsequent growth is largely basal and intercalary.[304] There is a sheathing basal or lower leaf region, and an upper leaf region comprising the 'blade'. During development growth is first concentrated in the upper leaf region, but later the rate of elongation increases in the lower leaf.

Adaxial growth first becomes conspicuous when the leaf is 30–40 μm long. Numerous radially aligned files of cells are established by rapid division in

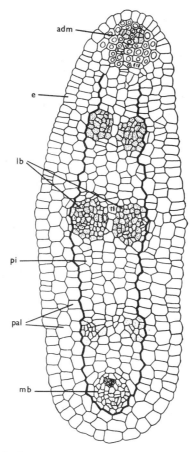

Fig. 5.11 T.S. developing phyllode of *Acacia longifolia*. (1020 μm high.) Activity of the adaxial meristem (adm) is about to cease. The shoot apex would be towards the top of the page, i.e. the phyllode is flattened in the radial plane of the shoot. e, epidermis; mb, median (i.e. abaxial) vascular bundle of the phyllode; lb, lateral bundles; pal, young palisade tissue; pi, central parenchyma. (From Boke,[47] Fig. 18, p. 82.)

the adaxial region. Gradually this localized meristematic activity is superseded by more diffuse intercalary growth. This very active adaxial meristem (Fig. 5.12), and subsequent intercalary meristematic activity on both sides of the centrally situated 'secondary midrib' region, result in considerable expansion of the leaf in the radial plane, so that it is radially, rather than dorsiventrally, flattened (Figs. 5.12 and 5.13). Procambial

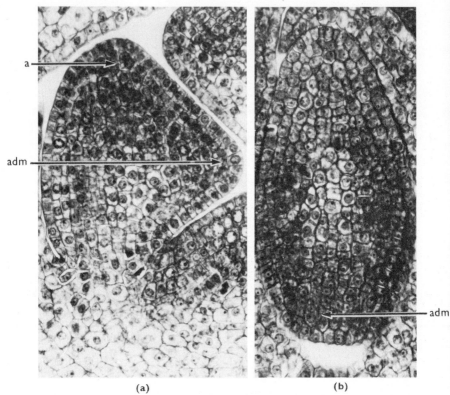

a

adm

(a)

adm

(b)

Fig. 5.12 *Acorus calamus*. (a) L.S. leaf primordium 120 μm long, showing extensive activity of the adaxial meristem (adm) and recently divided cells at the apex, a.
× 440. (b) T.S. leaf primordium 224 μm long, from the Wisconsin population. The
position of the shoot apex is towards the bottom of the page. The primordium is
extended in the radial plane, as a result of activity of the adaxial meristem (adm).
× 500. (From Kaplan,[304] Figs. 36 and 49, pp. 344 and 350.)

strands are differentiated in pairs on both sides of the leaf from derivatives
of intercalary meristems in abaxial and adaxial wings of the leaf.

Kaplan[304] made the interesting observation that leaves of *Acorus
calamus* collected from a site in Iowa differed in relative growth from those
collected in Wisconsin. Leaves from the Iowa population have more limited
radial growth (compare a and b in Figs. 5.14 and 5.15). Measurements of
radial diameter plotted against length of developing leaves from the two
populations demonstrate a progressive distinction in relative growth (Fig.
5.16a). That these differences are due to differences in cell number (i.e.
meristematic activity) and not to cell size can be shown by plotting the

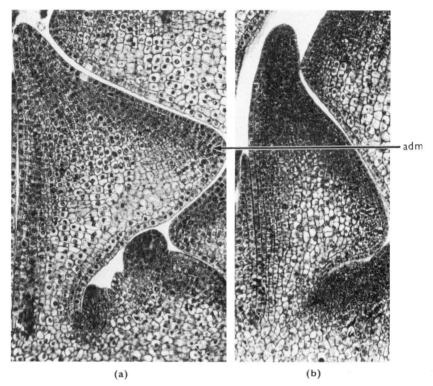

(a)   (b)

Fig. 5.13   Median L.S. leaf primordia of *Acorus calamus*. (a) Primordium 385 μm long from the Wisconsin population. The adaxial meristem (adm) has given rise to marked adaxial growth. (b) Primordium 370 μm long from the Iowa population, showing less adaxial growth. Both ×195. (From Kaplan,[304] Figs. 42 and 43, p. 346.)

number of cells along a single radial row against leaf length (Fig. 5.16b). In these geographical populations initial differences in activity of the adaxial meristem become enhanced during later development and result in measurable differences in the radial dimension. Correlated with this is the occurrence of only 6–7 pairs of vascular strands in leaves of the Iowa population, compared with 13 pairs in leaves of the Wisconsin population of the same length (Fig. 5.15).

In leaves of the Iowa population, not only is there less adaxial meristematic activity (Fig. 5.13a and b) but there is earlier and more extensive apical growth of the leaf than in those from Wisconsin. There appears to be

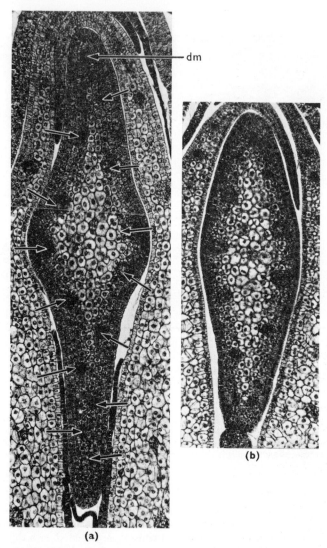

dm

(a)

(b)

Fig. 5.14 T.S. leaves of *Acorus* of equivalent lengths from the Wisconsin and Iowa populations. (a) From Wisconsin population. The dorsal median vein (dm) and several procambial strands (arrowed) are differentiating. Activity of a plate meristem is extending the leaf in the radial plane. (b) From Iowa population. The zone of maturing tissue is broader than that in the equivalent Wisconsin leaf. Procambial strands are differentiating. Compare Fig. 5.15. Both × 165. (From Kaplan,[304] Figs. 51 and 52, p. 350.)

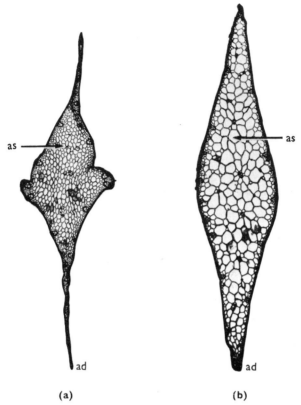

(a)   (b)

Fig. 5.15   T.S. mature leaves of *Acorus* from (a) Wisconsin, and (b) Iowa popu-lations. Conspicuous air spaces (as) are present. ad, adaxial position. Note the difference in shape of the mature leaves. Compare Fig. 5.14. Both ×6. (From Kaplan,[304] Figs. 53 and 54, p. 350.)

interaction between the radial and apical components of growth during early development of the leaf of *Acorus*. Although further work is required, present evidence suggests that the early differences in developmental pathways between leaves from the two localities are genetically determined rather than the result of factors in the environment.

Kaplan's[304] work on *Acorus* and that of Boke[47] on *Acacia* leave little doubt that the leaf of *Acorus* and the phyllode of *Acacia* are homologous organs. In both, marginal growth has been suppressed in favour of activity of an adaxial meristem, resulting in extension of the foliar organ in the

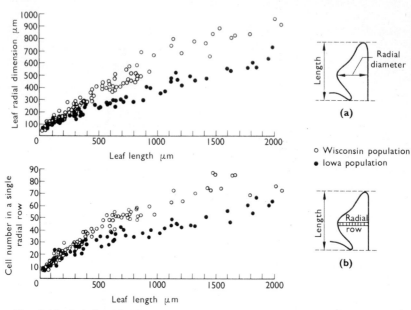

Fig. 5.16 Relationships between radial dimension, cell number and leaf length up to about 2000 μm (2 mm) in length from the Wisconsin and Iowa populations of *Acorus*. In (**b**) the relationship between cell number in a single radial row and leaf length is shown. (From Kaplan,[304] Fig. 62, p. 356.)

median plane. In *Acorus* radial extension proceeds at both edges and is bidirectional, in contrast to the unidirectional radial growth in *Acacia*. When leaves of *Acorus* are compared with those of other Araceae it is seen that adaxial and abaxial meristematic activity is accentuated in *Acorus* and only slightly expressed in the more conventional dorsiventral leaf forms in the family.[304]

Formation of cell layers: plate meristem

As a result of the activity of the marginal meristems a certain number of layers of mesophyll cells becomes established in the lamina at an early stage of development (Fig. 5.7). These cells then divide anticlinally and expand the lamina laterally (Figs. 5.7 and 5.8). Each cell divides to give a small 'plate' of cells, and the whole lamina functions as a so-called **plate meristem** (Fig. 5.2). This is usually considered to contribute to lateral growth of the leaf rather than to its thickness, but in *Xanthium* some periclinal divisions occur and form additional cell layers. In this plant experiments with radioactive isotopes indicated that the plate meristem is active for 3 days after marginal growth ceases, but both meristems are active for a total of about 23 days.[349] Few differences between the functioning of the marginal and plate meristems were observed.

An analysis of the planes of recent cell divisions at the edge of the lamina of leaves of *Xanthium* at three stages of development (Fig. 5.17) revealed

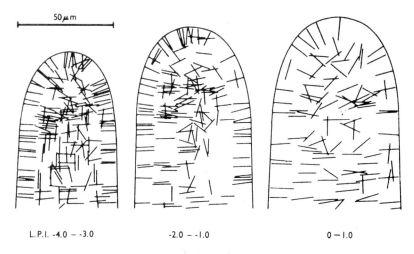

L.P.I. -4.0 − -3.0   -2.0 − -1.0   0 − 1.0

Fig. 5.17 Diagrams showing the orientation of cell plates in numerous T.S. margins of *Xanthium* leaves, at three stages as represented by the leaf plastochrone indices (L.P.I.) shown. The adaxial surface is to the left. (From Maksymowych and Erickson,[348] Fig. 5, p. 455.)

that divisions in the protoderm were predominantly anticlinal, i.e. at right angles to the surface. In the interior of the leaf the region within about four cell diameters of the margin could be considered as part of the marginal meristem, cells further removed from the margin being regarded as plate meristem. The preponderance in both regions of divisions at right angles to the horizontal plane is correlated with the lateral growth of the lamina.[348]

Usually the palisade tissue eventually differentiates from the adaxial layer of mesophyll formed by the marginal meristem, spongy mesophyll from the abaxial layer and sometimes parts of the middle layer, and the procambium of the veins from the middle layer or layers. However, as already mentioned, the layer which gives rise to the procambium is variable.[241]

In Fig. 5.18 the establishment and differentiation of these various tissues in the developing leaf of *Xanthium* are related to the length of the leaf and the thickness of the lamina. It is evident that most of the tissue layers are formed before the young leaf has attained much height.

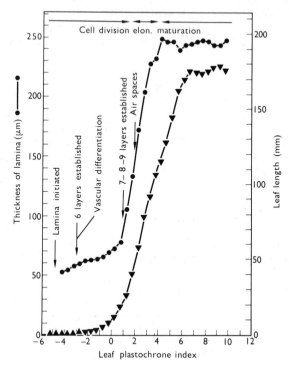

Fig. 5.18 Graph illustrating stages of development, as represented by changing leaf plastochrone index (L.P.I.), at which the processes indicated occur in leaves of *Xanthium*. (From Maksymowych and Erickson,[348] Fig. 9, p. 455.)

Development of compound leaves

In pinnate leaves, leaflet primordia arise marginally either in basipetal or acropetal sequence, or beginning more or less centrally and differentiating both acropetally and basipetally. Each leaflet follows a course of growth comparable to that of an entire leaf, i.e. exhibiting apical, intercalary and marginal growth. In compound leaves of this type marginal growth is essentially discontinuous, localized regions growing out as leaflets, or merely as lobes, and intervening loci undergoing less growth (in the case

Fig. 5.19 Developing leaf primordia of *Centaurea solstitialis*. (a)–(f) Primordia P_6–P_{11} from one shoot apex seen from the side, showing development of the lobes of the lamina, first evident in P_7. × 10. (g) A primordium from another apex. × 100. (h) A small mature leaf seen from above. × 1. l, lobes.

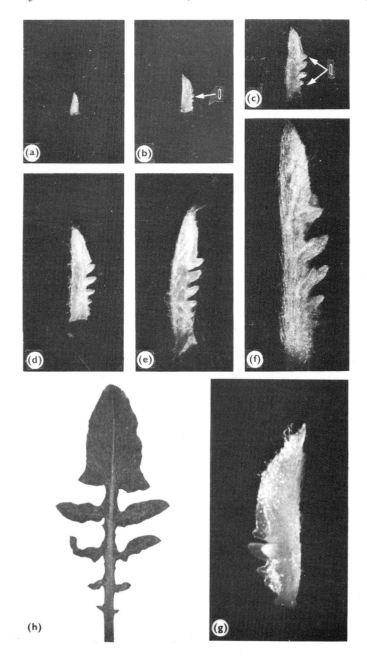

of lobed leaves) or none. This pattern of growth can be modified experimentally, as pointed out in a later section of this chapter.

The development of the lobed leaf of the yellow star thistle, *Centaurea solstitialis*, is illustrated in Fig. 5.19. The primordia of the lobes first appear as outgrowths of the margin of the young leaf primordium; in this species this is first visible in the 7th primordium from the apex, P_7 (Fig. 5.19b). The lobes are formed basipetally, older primordia exhibiting a greater number of lobe primordia (Fig. 5.19d–f). The development of a lobed leaf of this type is a consequence of differential activity of the leaf margin, some regions (the future lobes) showing considerable meristematic activity, while in other regions (the future sinuses, or inter-lobe regions) there is little or none. Procambial tissue is associated with the developing lobes. A fully developed leaf is shown in Fig. 5.19h.

CELL DIVISION AND EXPANSION

Both cell division and cell enlargement are involved in the growth of the young leaf. Cell division continues throughout a considerable part of the life of the leaf, until it is more than half or three-quarters grown, according to the species.[498] In lupin, about 90 per cent of the cells in the leaf are formed before it emerges from the bud;[498] in the lamina of the tobacco leaf this figure is about 62 per cent.[239] At this stage the tobacco leaf lamina is only about 28 per cent of its final volume, however, indicating that cell expansion will play an important role after emergence. Because about 99 per cent of the palisade cells in this species are produced after emergence of the leaf from the bud, and because prolonging cell division through one more cycle would have a profound effect on final leaf size after a substantial number of cells had been formed, it is sometimes argued that the events which take place after leaf emergence are more important in affecting leaf size than those that occur before emergence.[239] It seems likely that events before emergence may be of equal importance, however, since it is during this phase that the formation of a certain basic number of cells is achieved, and indeed in other species the rate of division prior to unfolding of the leaf is regarded as the most significant factor.[363] Premature emergence, or expansion, of the leaf results in leaves of small size, as in many seedlings. These matters are of considerable practical importance, since yields of many crops depend on the total leaf area attained by the plant. The rate of growth of the leaf may be controlled by the older leaves, as in banana.[31]

During the growth of the leaf, the cell number doubles itself many times. Early in the development of the tobacco leaf the time required for the cells to double in number is only 0·4 days;[239] in *Xanthium* this 'cell generation time' is approximately 2·2 days.[345] During the formation of the lupin leaf about 13 or 14 cell generations are involved, compared with 11 or 12 in sunflower[498] and 27 in *Xanthium*.[345] Thus there is a good deal of variation in different species.

In an excellent analysis of leaf development in clover, *Trifolium repens*, Denne[139] observed that the relative growth rate was highest when first measurable after leaf inception, and that it then declined. In tobacco, also, the relative growth rate declined.[239]

Using the shoot of *Xanthium*, Erickson and Michelini[163] derived a developmental measure, the **plastochrone index**. The plastochrone serves as the unit of the developmental scale, and the index is linearly related to time. The leaf plastochrone index (L.P.I.) of a leaf of *Xanthium* 10 mm long is zero, leaf primordia shorter than 10 mm having a negative L.P.I. and those longer than 10 mm a positive one. During early stages of development in the *Xanthium* leaf, which is one of the most thoroughly investigated angiosperm leaves, the number of cells increases exponentially (Fig. 5.20). At about L.P.I. 3.0 (equivalent to a leaf length of 76 mm) the curve levels off, indicating that the cells of the lamina have ceased dividing. It is important to note that not all the regions and tissues of the young leaf grow at the same rate. For example, in *Xanthium* the lamina stops growing about 1·5 plastochrones (*c.* 6·6 days) before the petiole, and, as in other species, the tip of the leaf ceases cell division and growth and becomes mature about two plastochrones (*c.* 8·8 days) sooner than the basal part.[345, 346] Within the lamina itself, growth of the various immature tissues is not uniform, especially cell expansion, although studies with ³H-thymidine also show differential labelling, suggesting that DNA synthesis, and mitosis, occur at different rates. Thus in young developing leaves of *Xanthium* supplied with ³H-thymidine histoautoradiographs showed that, as might be expected, the amount of DNA synthesis decreased with the age of the leaf. In the cells of the palisade layer, DNA synthesis was constant between a L.P.I. of −0·56 and 0·94 (corresponding to lengths of about 7·0 and 22·5 mm), but less than in the other tissues; between L.P.I. of 0·94 and 2·2 (*c.* 22·5–55·0 mm length) there was an increase, and the difference in the amount of DNA synthesis between the palisade cells and those of other leaf tissues was highly significant. Synthesis continued longer (at least one plastochrone) and at a significantly higher rate in the palisade layer than in the epidermal and mesophyll cells.[347] In general, cell division usually continues longer in the palisade than elsewhere in the leaf. In clover, cell division ceased first in the lower epidermis, then upper epidermis and mesophyll, and lastly in the palisade layer (Fig. 5.21).[139]

During the growth of the leaf the cells enlarge considerably; for example, in lupin the cell volume increased 40-fold, and in sunflower 90-fold, over a period of time.[498] In *Xanthium* the most rapid rate of cell expansion occurs when the leaf is about 44 mm long, when most cell division has ceased.[348]

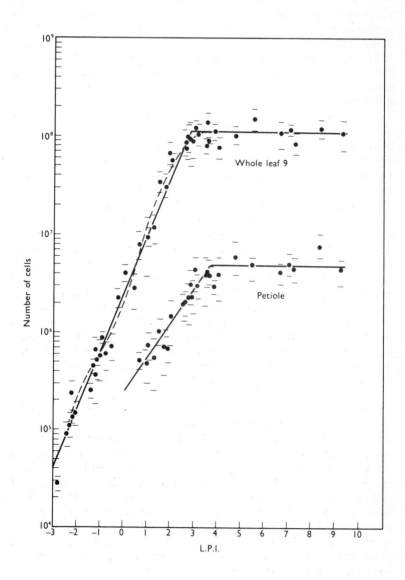

Fig. 5.20 Number of cells in the whole leaf and in the petiole of leaf 9 of *Xanthium* (numbered in order of appearance on the shoot) at different stages in its development, as represented by leaf plastochrone index (L.P.I.). (From Maksymowych,[346] Fig. 1, p. 896.)

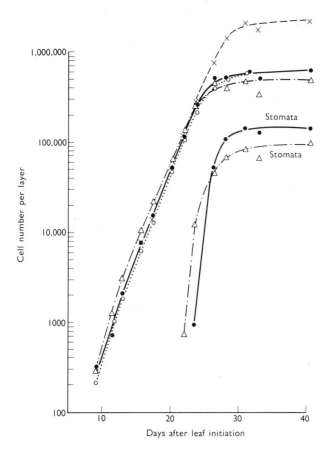

Fig. 5.21 Mean cell number per layer in developing leaflets of clover, *Trifolium repens*. All points represent the mean of at least 3 leaflets. Numbers of stomata are not included in the number of epidermal cells but are shown separately. ●—●, upper epidermis; ×——×, palisade; ○···○, mesophyll; △—·—△, lower epidermis. (From Denne,[139] Fig. 3a, p. 206. © 1966 by the University of Chicago. All rights reserved.)

Cells of the epidermis usually continue to enlarge for a period after the mesophyll cells have reached their final size. There may be both differential rates and duration of cell expansion in the different tissues. In *Xanthium* the palisade cells increased their volume 11-fold during growth, while the cells of the upper epidermis underwent a 29-fold increase.[346] In his classic study of the development of the tobacco leaf, Avery[20] pointed out that the cells of the epidermis continued to enlarge after the cells of the middle and abaxial mesophyll had ceased expansion. As a consequence of the forces thus set up, the mesophyll cells are pulled apart, resulting in the branched cells and numerous intercellular spaces which comprise the spongy meso-phyll. Avery considers that the cells of the lower epidermis are subject to a reciprocal pull which distorts the anticlinal walls and makes them appear sinuous. The palisade tissue may be pulled apart slightly also, and in turn its stresses may lead to the less marked waviness of the lateral walls of the upper epidermis.

Some leaves, notably those of *Monstera* and the aquatic plant *Aponogeton*, have holes in the lamina. These develop by necrosis of patches of tissue during the ontogeny of the leaf. In *Monstera*, pairs of spots are equally spaced within the tissue delimited by each pair of lateral veins. The distances between the pairs of necrotic spots is constant, and it is sugges-ted[357] that their regular distribution can be explained in terms of Turing's[532] diffusion-reaction theory of morphogenesis.

Differential rates and durations of cell division and cell expansion are thus very important in affecting the structure of the leaf, an organ highly specialized for photosynthesis and hence important in food production. It is consequently surprising that rather little is known about the control of these aspects of growth in the leaf, complex as they are.

STRUCTURE OF THE MATURE LEAF

At maturity a dorsiventral foliage leaf of a representative mesophytic dicotyledon is made up of upper and lower **epidermis, ground tissue (mesophyll)**, which forms the main photosynthetic region, and **vascular tissue**, forming the **veins**. In transverse section these tissues are seen, from top to bottom, as the upper or adaxial epidermis (the side towards the axis), a region of parenchymatous cells elongated in the transverse plane of the leaf and containing numerous chloroplasts, the palisade tissue, then a region of irregularly shaped cells with sparser chloroplasts and numerous conspicuous intercellular spaces, the spongy mesophyll, and the lower or abaxial (away from the axis) epidermis (Fig. 5.22). Vascular strands are present in the central region of the leaf, oriented in such a way that the phloem is abaxial and the xylem adaxial; this orientation is a consequence of the continuity of vascular tissue between leaf and stem.

Fig. 5.22 T.S. dorsiventral leaf of *Syringa*. **(a)** ×150. **(b)** ×300. e, upper epidermis; g, guard cells of a stoma in the lower (abaxial) epidermis; h, glandular hair; p, palisade; s, spongy mesophyll; sc, sub-stomatal chamber.

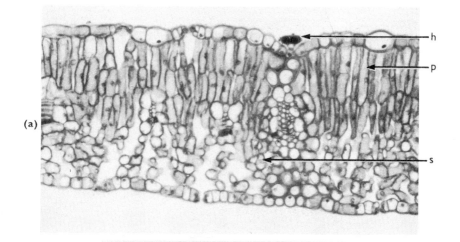

(a)

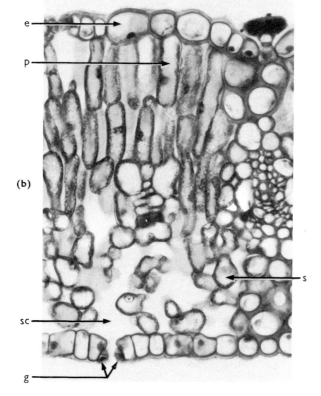

(b)

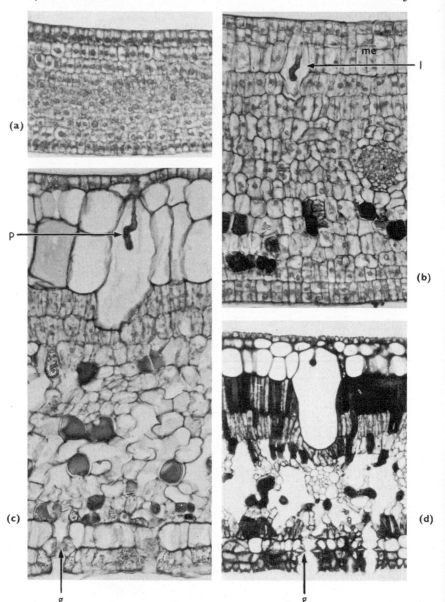

Fig. 5.23   T.S. leaf of *Ficus* at different stages of development. (**a**) Young leaf in which the protoderm has not yet divided to give the multiple epidermis.   × 300. (**b**) Slightly older leaf, in which the cells of the upper and lower epidermis have

Epidermis

Usually a single layer of epidermal cells is present, but the leaves of some plants have a multiple epidermis, e.g. *Ficus, Nerium, Piper* (Figs. 5.23–5.25). In such leaves the cells of the protoderm divide periclinally to give rise to additional layers (Fig. 5.23a–c). The number of layers of cells varies from 2 to about 16, according to the species; the multiple layers probably protect the mesophyll from undue desiccation. The adaxial epidermis may have more layers than the abaxial epidermis of the same leaf. Where the abaxial epidermis has several layers, a considerable sub-stomatal cavity is usually present, forming a space between the guard cells and the underlying photosynthetic tissue (Fig. 5.25). In the leaf of *Nerium*, the oleander, stomata are restricted to indentations of the abaxial surface, the stomatal crypts; in these regions the epidermis is only a single layer (Fig. 5.24).

Stomata may occur on both sides of the leaf, or only on one, more commonly the abaxial surface; in floating leaves, e.g. those of the water lily, *Nymphaea*, they occur only on the adaxial surface. Stomata may be level with the other epidermal cells, raised above this level, or more or less sunken (Fig. 5.35); they may also be restricted to certain loci in the leaf, such as the stomatal crypts of *Nerium* mentioned above.

The development, structure and distribution of stomata are discussed more fully in Part 1,[127] Chapter 7. Since, however, stomatal development is a good example of the general process of differentiation, and since some recent work with the electron microscope has been reported, some discussion is included here. In grasses and other monocotyledons the guard cell mother cell is formed by the unequal division of a protodermal cell. The guard cell mother cell is the smaller, more highly cytoplasmic product of this division (Figs. 5.26 and 5.27). Small subsidiary cells are formed adjacent to the guard cell mother cell by the division of adjacent epidermal cells (Figs. 5.26c–f and 5.30), and later the guard cell mother cell divides to give rise to the two guard cells (Figs. 5.26e and f, and 5.31). Stages in this process in the leaf of *Cyperus alternifolius* and in the internodal epidermis of *Avena*, oat, are shown in Figs. 5.26 and 5.28–5.32. Recent studies of the epidermis in developing leaves of wheat with the electron microscope have shown that before the unequal division the nucleus becomes displaced

---

divided periclinally to give several layers of cells. The lithocysts (l) have not divided. × 300. (c) Older stage, in which sunken guard cells have differentiated in the abaxial epidermis. Two layers of developing palisade tissue are evident. The peg of wall material (p) in the lithocyst on which a cystolith of calcium carbonate is deposited can be seen; the cystolith itself has been almost totally dissolved in the preparation of the section.   × 300. (d) Mature leaf, showing the sunken position of the guard cells.   × 150. g, guard cells; l, lithocyst; me, multiple epidermis; p, peg of cell wall.

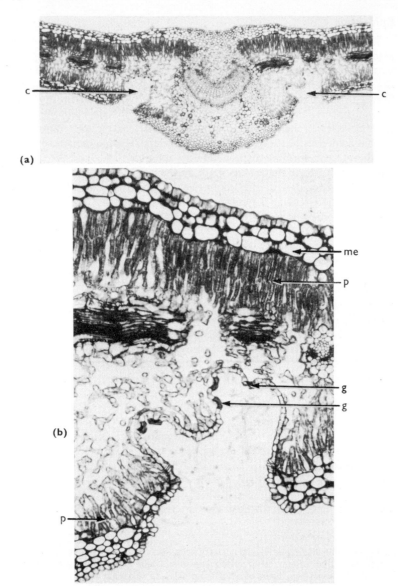

Fig. 5.24 T.S. leaf of *Nerium oleander*. (**a**) Part of midrib region of the isobilateral leaf.   × 40. (**b**) Part of the lamina in the region of a stomatal crypt, showing the raised guard cells (g) restricted to these regions.   × 300. c, stomatal crypts; g, guard cells; me, multiple epidermis; p, palisade on both sides of the leaf.

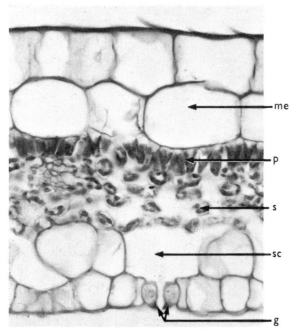

Fig. 5.25   T.S. leaf of *Piper* sp. A multiple epidermis (me) is present on both sides of the leaf. A large sub-stomatal chamber (sc) is present between the guard cells (g) and the spongy mesophyll (s). p, palisade.   × 300.

towards one end of the cell, and vacuoles to the other end, other organelles being uniformly distributed. Before prophase, a band of microtubules was evident round the nucleus.[406] After formation of the guard cell mother cell, the nuclei of adjacent epidermal cells move towards the guard cell mother cell and divide to form the subsidiary cells. A hemispherical cell plate forms, apparently aligned by microtubules. Subsequently the guard cell mother cell divides symmetrically to give rise to the guard cells, the intervening wall being incomplete with one or more holes at either end.[406] Similar events occur during the formation of stomata in the internodal epidermis of oat, and again there are well defined openings between the two guard cells.[307] Stages in the formation of guard cells in the stem epidermis of oat are illustrated in Figs. 5.28 to 5.32. Although the tissue investigated was the epidermis of the developing stem, the process is essentially similar to that taking place in young leaves. The movement of the nucleus to one end of the cell prior to the unequal division leading to formation of the guard cell mother cell is evident (Fig. 5.28), as is the concentration of

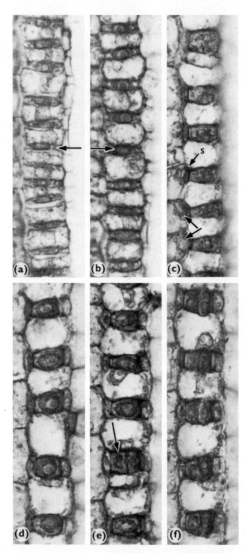

Fig. 5.26 Stages in the development of the stomatal complex as seen in para-
dermal sections of the young leaf of *Cyperus alternifolius*. The shoot apex is
towards the bottom of the page. (a)–(f) represent increasingly distal strips of epidermis. In (a) and (b) the small guard cell mother cells are arrowed. Intervening
epidermal cells are much more vacuolate. In (c) nuclei of adjacent epidermal cells
(arrowed) have moved close to the guard cell mother cell prior to dividing to form
subsidiary cells (s). (d) shows a stage after formation of subsidiary cells but before

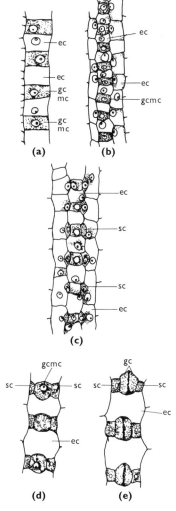

Fig. 5.27 Stages in development of the stomatal complex in leaves of rice, *Oryza sativa*. ec, epidermal cell; gc, guard cell; gcmc, guard cell mother cell; sc, subsidiary cell. Use as a key to Fig. 5.26. (From Kaufman,[305] Figs. 73–76, p. 306.)

the division of the guard cell mother cell. In (e) a guard cell mother cell (arrowed) has divided to form two guard cells. In (f) all stomatal complexes show 2 guard cells and subsidiary cells. All × 1000. Use Fig. 5.27 as a key to this figure. (Slide by courtesy of Dr. J. B. Fisher.)

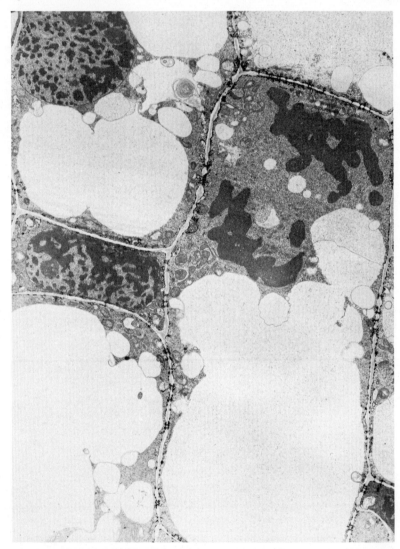

Fig. 5.28 Unequal division of an epidermal cell in the developing internode of oat, *Avena sativa*, in the intercalary meristem region. Prior to mitosis the nucleus and much of the cytoplasm has moved to the end of the cell (on the right). Mitosis is in progress. On the left a small densely cytoplasmic cell and a large, highly vacuolate cell, the products of such an unequal division, are seen. The small cell will probably develop as a guard cell mother cell (in this region, however, some small cells develop in other ways, e.g. as trichomes). × 1880. (By courtesy of Dr. P. B. Kaufman.)

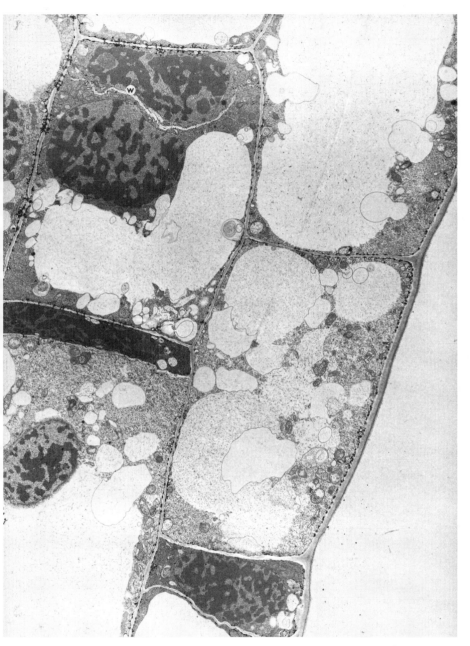

Fig. 5.29 Short and long cells in the meristematic region of the internodal epidermis of oat. Note the dense cytoplasm of the short cells, presumptive guard cell mother cells, and the much greater degree of vacuolation of the long cells. The wall (w) between the guard cell mother cell and its sister cell is markedly curved. × c. 3800. (By courtesy of Dr. P. B. Kaufman.)

6

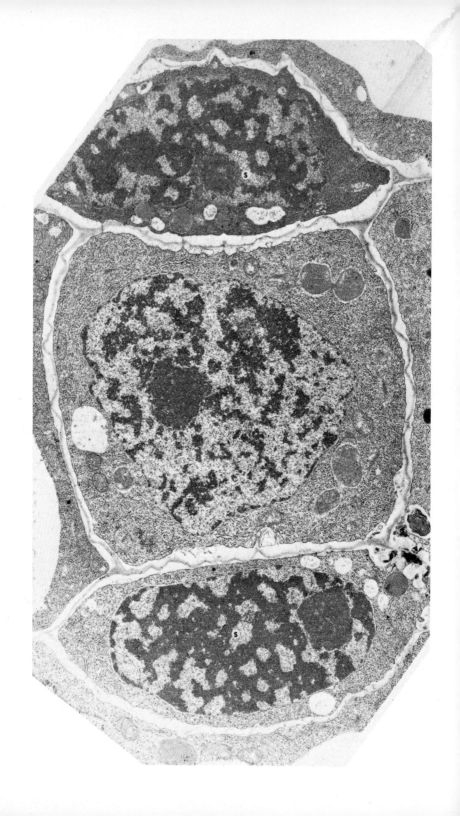

cytoplasm at that end of the cell and in the smaller product of the unequal division (Fig. 5.29). The larger product of the division, which develops as an ordinary cell of the epidermis, is much more highly vacuolate. These illustrations afford a striking example of polarized cellular differentiation in a single layer of tissue. Figs. 5.28 and 5.29 may be compared with Figs. 2.9 and 2.10, illustrating the unequal division of cells in the root epidermis in the formation of trichoblasts. The contrast between the larger and smaller product of the unequal division, both in size and density of cell contents, is greater in the guard cell mother cells illustrated than in the example of trichoblast formation.

After the formation of the two guard cells in oat, the nuclei become much elongated and plastids with a few grana become evident. A pore develops between the two guard cells (Fig. 5.32). Microtubules become apparent along the common wall between the guard cells, and the wall forms a thick pad in this region.

In dicotyledons such as *Dianthus*, the guard cell mother cells also develop from the smaller product of an unequal division,[393] but it is more difficult to follow their development in detail. In grass leaves the stomata develop sequentially in longitudinal rows, because of the activity of the basal intercalary meristem, but stomata of dicotyledonous plants are more scattered. Recently, however, studies have been carried out on the ontogeny of stomata in a number of dicotyledonous groups.[264–267] Ontogeny is often somewhat variable even in a single species.

Epidermal cells do not contain well developed plastids, with the exception of the guard cells, though even here the plastids have few grana. In some aquatic species, however, as for example *Phyllospadix* which grows immersed in sea water, chloroplasts do occur in the epidermal cells. An excellent example is to be seen in the epidermal cells of the submerged leaf of *Ranunculus*, illustrated in Fig. 5.43.

Both covering and glandular **hairs**, or trichomes, may occur on the leaf epidermis (Fig. 5.8c). They show a considerable range of form (see Part 1,[127] Chapter 7). Cells containing crystals, usually cystoliths of calcium carbonate, may also occur as idioblasts in the leaf epidermis. These cells, though without the crystals, are shown in Fig. 5.23b–d.

Fig. 5.30 Guard cell mother cell with two subsidiary cells (s) from internodal epidermis of oat. The orientation of this figure and of Fig. 5.31 is at right angles to the orientation of Fig. 5.26–5.29. × c. 3500. (By courtesy of Dr. P. B. Kaufman.)

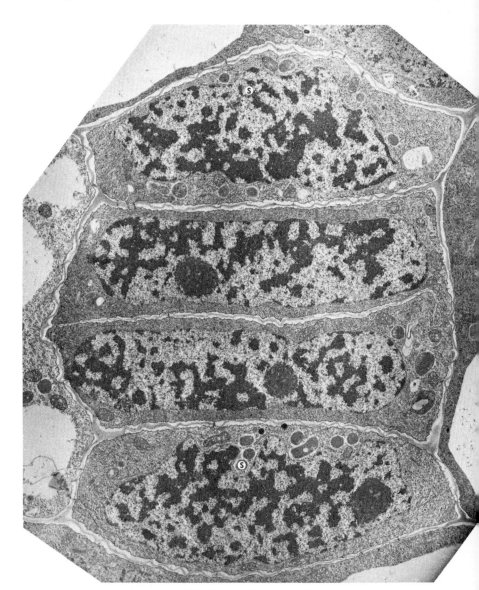

Fig. 5.31 Early stage in development of stomata in the internodal epidermis of oat. The guard cell mother cell has divided by a wall parallel to the long axis of the stem to give the guard cells. s, subsidiary cell.   × 7500. (From Kaufman *et al.*,[307] Fig. 16, p. 39.)

The epidermal cell walls may occasionally contain lignin, and more frequently cutin; also a superficial layer of cutin, the cuticle, is often present. The cuticle varies in thickness and in leaves of some xerophytes may be quite thick.

A recent study of the leaves of garden beet, *Beta vulgaris*, disclosed some interesting effects of exposure to air pollution on the leaf surface.[76] Exposure to vehicle exhaust fumes, or merely exposure of the plants for 7 days or even 24 hours to heavy smog in Los Angeles, resulted in changes in the pattern of extruded wax on the leaf surface. Excessive extrusion took place in irregular spots, the wax deposits occurring in various formations, as opposed to a more moderate and uniform extrusion of wax in leaves of control plants kept in a smog-free greenhouse. The changes were related to changes in the rate of extrusion of rodlets of wax. It is interesting to speculate upon the physiological changes which may accompany this morphological manifestation of smog damage.

Another investigation showed that in plants of *Nicotiana glutinosa* exposed for only 2 hours to Los Angeles smog there was damage to leaves at certain stages of development, notably those undergoing expansion.[213] Damage was localized and was associated with cells which had just attained maximum size. It occurred in regions of the leaves where stomata had just become functional and where, accordingly, polluted air could more readily penetrate to the interior tissues through sub-stomatal cavities and inter-cellular spaces in the mesophyll.

Mesophyll

In most dicotyledons, the ground tissue, or mesophyll, is differentiated into *palisade* and *spongy mesophyll*. Cells of the palisade are cylindrical, elongated in the transverse plane of the leaf, contain many chloroplasts and are densely packed together into one or more layers (Fig. 5.22). In dorsiventral leaves the palisade tissue occurs on the adaxial surface only; in isobilateral leaves, palisade cells occur on both surfaces of the leaf but may be less well developed on the abaxial side. In cylindrical leaves, such as those of some species of *Hakea*, the palisade tissue occurs all round the periphery of the leaf (Fig. 5.34). At maturity two layers are present (Fig. 5.35); osteosclereids occur as elongated idioblasts the length of which is equivalent to the depth of both palisade layers. Although this leaf is cylindrical at maturity, it is dorsiventral near the base in early stages of development (Fig. 5.33).

The spongy mesophyll, where present, consists of irregular, branching cells separated by large air spaces and containing chloroplasts (Fig. 5.22). There is a greater volume of intercellular spaces than of cells.

Mesophyll cells of a number of species have been successfully isolated and cultured *in vitro*. Long ago Haberlandt predicted that parenchyma

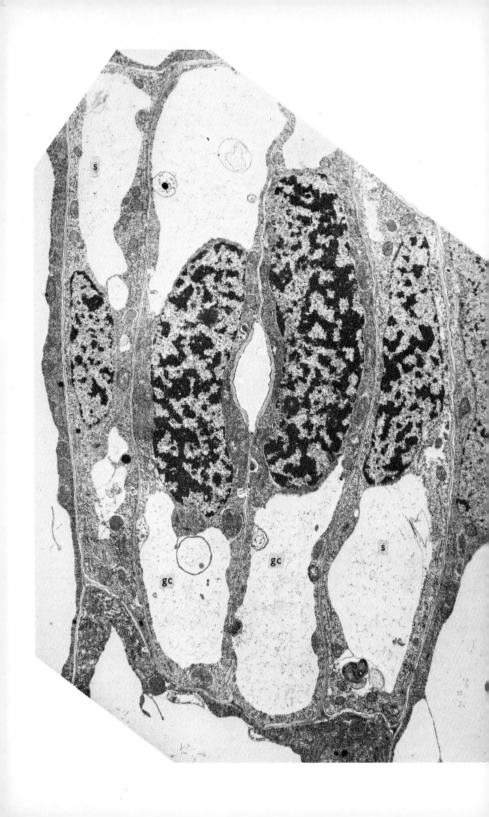

cells might be capable of developing into whole plants, and considered that the mesophyll tissue, with its relatively loosely organized structure, might be suitable for this. Recently it has been shown[294] that both palisade and spongy mesophyll cells can divide in culture. In these experiments a single cell could divide to give a group of 20–30 cells in a period of 14 days. Culture of palisade cells was much more successful than of the spongy mesophyll, perhaps because, as we have seen, DNA synthesis normally persists longer in the palisade tissue. Roots were sometimes formed from the spheres of tissue resulting from division of the mesophyll cells.[295] Thus relatively mature cells of the leaf mesophyll are capable of further growth and development if removed from their normal environment.

In some leaves, for example those of grasses, the mesophyll is not differentiated into palisade and spongy tissue but comprises parenchyma cells of fairly uniform size and structure (Fig. 5.36b). In grasses, and in many other monocotyledonous and dicotyledonous plants, the cells

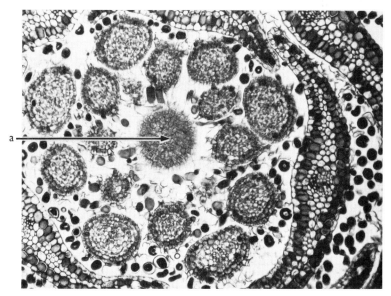

Fig. 5.33 T.S. shoot apex of *Hakea constablei*. Primordia of foliage leaves surround the shoot apex (a); thinner, flattened bud scales are present on the outside of the bud. The foliage leaves are radially symmetrical in the upper parts but dorsiventral near the base.   × 150.

Fig. 5.32 Stomatal complex in the internodal epidermis of oat. The guard cells (gc) and subsidiary cells (s) are now somewhat vacuolate. A pore has formed between the guard cells.   × 6400. (From Kaufman *et al.*,[307] Fig. 23, p. 44.)

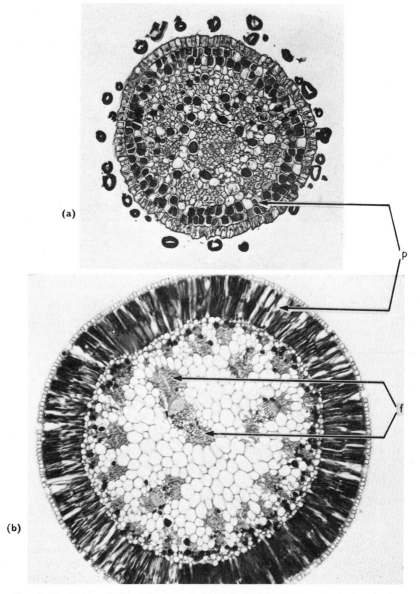

Fig. 5.34 T.S. cylindrical leaves of *Hakea* sp. (**a**) Young developing leaf. Palisade cells (p) are not yet elongated.   × 150. (**b**) Older leaf, showing two rows of elongated palisade cells all round the periphery of the leaf and numerous sclerenchymatous fibres (f) on both sides of the vascular bundles.   × 50.

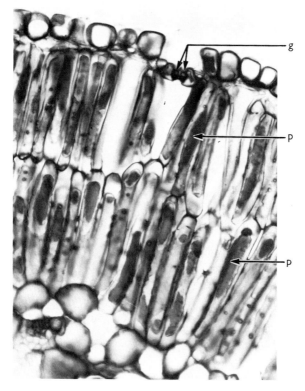

Fig. 5.35   T.S. mature leaf of *Hakea* sp. with two rows of columnar palisade cells (p) and guard cells (g) and subsidiary cells of the stomata sunken below the level of the epidermis.   × 300.

surrounding the vascular tissue are morphologically different from adjacent mesophyll cells. They are often larger, may have fewer chloroplasts, and may be more thick-walled. These cells constitute the **bundle sheath** (Fig. 5.36a, b); in some species they extend to the leaf surfaces, forming the bundle sheath extensions. There is some evidence that the bundle sheath and its extensions function in conduction and in storing food materials. The bundle sheath can be considered as an endodermis, since Casparian strips are occasionally observed, or as a starch sheath, since starch may occur in these cells. Recent work indicates considerable structural and functional specialization of the bundle sheath cells, at least in some species.

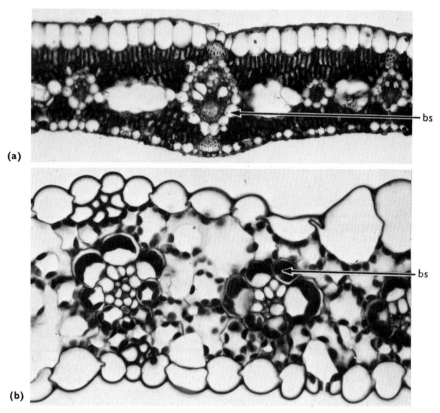

(a)

(b)

Fig. 5.36   (a) Hand section of a fresh leaf of *Cyperus alternifolius*. The cells of the bundle sheath (bs) are quite different in appearance and plastid complement from those of the adjacent mesophyll.   × 150. (b) T.S. part of the leaf of *Zea mays*. Bundle sheath cells (bs) have much larger plastids (here the cells are somewhat plasmolysed) than adjacent mesophyll cells.   × 500.

Structure and function of bundle sheath cells

Recent work with the electron microscope on plants with particular biochemical properties has suggested a very interesting function for the structurally specialized cells of the bundle sheath. Much of this work has been done with maize, sugar cane and other tropical grasses.

It has recently been shown[243] that an alternative pathway of photosynthetic carbon fixation to the Calvin cycle exists in certain tropical grasses. This C_4-dicarboxylic acid pathway, known as the Hatch-Slack pathway, is characterized by early labelling with $^{14}CO_2$ of the C_4 positions of 4-carbon compounds in contrast to the early labelling of 3-carbon compounds in the Calvin cycle. Some tropical grasses are capable of lowering the concentration of CO_2 in a closed system to less than 5 p.p.m., and are therefore said to show low CO_2 compensation values.[151] Most plants have relatively higher compensation values, being capable of lowering the concentration to only about 50 p.p.m., and they produce CO_2 in the light by a process often referred to as photorespiration. Activity of the enzyme phosphoenolpyruvate carboxylase was about 60 times greater in the tropical grasses than in other species tested, and is believed to be the principal photosynthetic carbon dioxide-fixing enzyme in these grasses.[474] Most of the enzyme was found in the chloroplast fraction of extracts, and is presumably associated with the plastids.

All grasses belonging to the chloridoid-eragrostoid and panicoid taxonomic divisions of the Gramineae have low CO_2 compensation values, except one genus which is believed on other grounds to be taxonomically misplaced. These grasses also have certain anatomical features in common, namely specialized bundle sheath cells in the leaves. Some dicotyledons, members of the Amaranthaceae and Chenopodiaceae, also carry out photosynthesis by the Hatch-Slack pathway. The leaves of these species, too, have prominent bundle sheaths, surrounded by a layer of palisade cells.[151, 319] Starch is present in the plastids of the bundle sheath cells, whereas in species with high CO_2 compensation values starch occurs throughout the mesophyll.[151]

Studies with the electron microscope have shown that the chloroplasts in the bundle sheath cells of *Amaranthus edulis* and *Atriplex lentiformis* are large, have grana and contain abundant starch grains; those in the palisade cells are smaller and contain little starch (Fig. 5.37).[319] They thus resemble those of tropical grasses, except that at maturity the chloroplasts of the bundle sheath cells of the latter have no grana.[259, 321] Apparently in these species with very high rates of photosynthesis the bundle sheath cells are adapted to receive photosynthetic products from the mesophyll cells which are radially arranged round them, and to store them temporarily as starch. The mitochondria of the bundle sheath cells, which are much larger than those of adjacent mesophyll cells, may provide some of the energy for this transport system of photosynthetic products.[319] The bundle sheath cells are metabolically much more active than those of the surrounding mesophyll.[45] In the chloroplasts of maize, *Zea mays*, the internal membrane system is continuous with the plastid envelope.[432] This type of plastid structure, in which there is more extensive direct connection between the loculi, fret channels and the inner part of the plastid envelope, is apparently found only in plants having the Hatch-Slack pathway of carbon dioxide fixation.

It is of interest that the species which show these physiological and

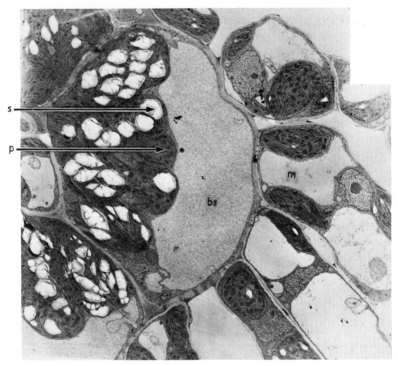

Fig. 5.37 Electron micrograph of part of the bundle sheath (bs) and mesophyll (m) cells of the leaf of *Amaranthus edulis*. The plastids (p) of the bundle sheath cells contain abundant starch (s); those of the mesophyll cells do not. Both types of plastid have grana.   × 3000. (By courtesy of Dr. W. M. Laetsch.)

structural features probably all originated in the tropics. It is believed that these features may have evolved in response to particular environmental factors;[151] one such factor might be high light intensity under conditions where water stress might often be a factor limiting photosynthesis.[319]

It has been suggested[45] that the proximity of the photosynthetic tissues to the water-conducting tissues should offer a protective advantage to species subjected to desiccation. Tropical grasses and some other species are capable of net rates of photosynthesis that are about double those found in plants with photorespiration. It has been stated that species from temperate regions, with chloroplasts diffused throughout the leaf meso-phyll, should saturate at a lower incident density, whereas a much higher light requirement should be expected for tropical grasses because of the density of chlorophyllous tissue around the vascular bundles.

An interesting ontogenetic study of the sugar cane leaf[320] has shown that the plastids of both bundle sheath and mesophyll cells, eventually so different in structure and function, both originate from proplastids which are not morphologically distinguishable. During development the chloroplasts in both types of cell develop grana, but those in the mesophyll have more thylakoids in the grana and also have well developed prolamellar bodies. During the terminal phase of chloroplast growth the plastids in the bundle sheath cells lose their grana; at maturity they have no grana but can accumulate large amounts of starch (Fig. 5.38). Thus in sugar cane the structure of the chloroplasts in the specialized bundle sheath cells is a result of reduction;[320] in maize the chloroplasts in the bundle sheath cells do not lose all their grana. Presumably biochemical conditions in the bundle

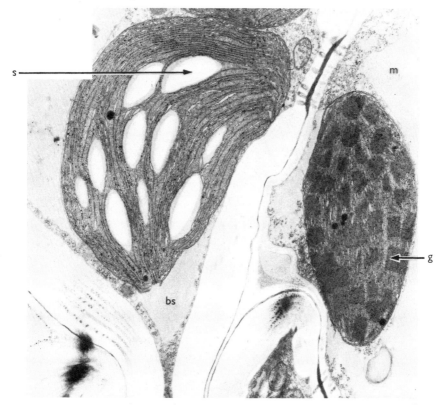

Fig. 5.38 Plastids in the bundle sheath (bs) and mesophyll (m) cells of the leaf of sugar cane, *Saccharum officinarum*. Plastids in the mesophyll cells have evident grana (g), those in the bundle sheath cells lack grana but store starch grains (s). × 12,300. (From Laetsch,[319 a] Fig. 5, p. 329.)

sheath and mesophyll cells differ sufficiently to influence plastid development.

It seems likely that this remarkable example of the relationship between structure and function, with the development of bundle sheath cells containing specialized organelles and having a very specific function, may be a long-term response to evolutionary forces in the tropical environment.

Vascular tissues

One of the more evident morphological features of the mature leaf is its *venation*, a complex system of vascular bundles. Leaves of most dicotyledons have one main vascular bundle in the midrib and a network of progressively smaller veins, forming a reticulate venation system (Fig. 5.39). Most monocotyledons, on the other hand, have a parallel venation

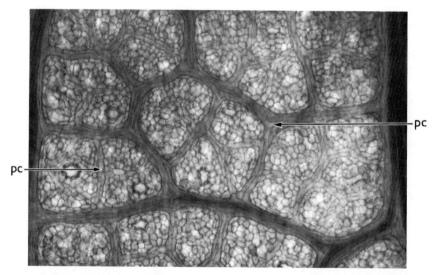

Fig. 5.39 Network of procambium (pc) between secondary veins in a young leaf of *Liriodendron* with a lamina 3 mm long. The material has been cleared and stained. × 215. (From Pray,[411] Fig. 7, p. 23.)

system, in which several more or less equal vascular bundles are connected by smaller commissural veins running between the major veins. Many descriptive studies of venation exist, but little is known of the factors controlling the development of the venation systems.

It is considered that during evolution reticulate venation patterns probably originated from open dichotomous systems of venation such as occur in various lower groups of Pteropsida.[192] These are bifurcating systems in which the forks end blindly without joining other strands. Open dichotomous venation is found in a few living angiosperms, such as *Kingdonia*[194] and *Circaeaster*.[193] Anastomoses between veins in these species are very infrequent.

In most dicotyledons a central procambial strand differentiates acropetally into the leaf primordium and eventually forms the vascular tissue of the midrib. It is continuous with vascular tissue in the stem. At a later stage in development, phloem differentiates acropetally in the procambial strand; slightly later, xylem elements usually differentiate at about the level of attachment of the leaf primordium to the axis, and subsequently develop both acropetally and basipetally. Thus the xylem is initially discontinuous. Jacobs and Morrow[271] have shown that, in *Coleus* at least, there may be more than one locus of xylem differentiation and more than one region of discontinuity. In most species, phloem is present abaxially and xylem adaxially in the midrib bundle, in continuity with these tissues in the stem (Fig. 5.7d). In species with bicollateral vascular bundles in the stem, the internal phloem may continue into the leaf, so that phloem may occur adaxial as well as abaxial to the xylem. This usually is present only in the major vascular bundles of the leaf.

Secondary veins differentiate acropetally from part of the ground tissue region of the leaf lamina. The cell layer from which they differentiate varies, as already discussed. In some species, a marginal vein differentiates before the secondary veins, which extend between the midrib and marginal veins.[421] In the leaf of *Liriodendron*, the tulip tree, for example, the secondary veins consist from the outset of several layers of cells; they differentiate progressively towards the leaf margin. Tertiary and then quaternary veins later develop between them, forming the familiar complex network of veins characteristic of the adult leaf (Fig. 5.39). The commissural tertiary veins which link up the secondaries apparently all differentiate simultaneously.[411] In the differentiation of a minor vein a single series of cells becomes conspicuously enlarged and then differentiates into procambium. Following the differentiation of the minor veins, regions of mesophyll may be completely enclosed by vascular strands; such regions are called vein islets. In other regions a vein may end blindly in the mesophyll; such vein endings are usually composed of both xylem and phloem, although sometimes only a single sieve element is present, surrounded by xylem.[169]

In broad-leaved monocotyledons, such as *Hosta*[412] and the water hyacinth, *Eichhornia*, a number of separate primary veins are present; these fan out in broad arcs and join as they approach the tip of the leaf. Commissural or intercostal veins run transversely between the primary veins, and also form loops and curves, resulting in a complex network. In *Hosta* the commissural veins differentiate from a layer of narrow elongated cells, formed by precisely oriented cell divisions, running perpendicular to the

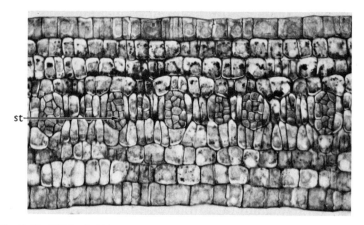

Fig. 5.40   T.S. leaf of *Hosta* taken parallel to the primary veins. Commissural strands (st) are shown in section. Small veins occur between the larger ones.   × 385. (From Pray,[413] Fig. 7, p. 703.)

direction of the primary veins (Fig. 5.40).[413] These veins differentiate first near the tip, and lastly near the base of the leaf, but additional strands may differentiate between the first-formed ones.

Whereas in these leaves procambium is continuous with existing procambium, in the leaf primordia of wheat and a number of herbage grasses the procambial strands which form the primary veins differentiate at the insertion of the primordium on the axis but independently of the rest of the vascular system, with which they are not initially continuous.[258, 465] In wheat successive primary strands originate in the primordium itself, and differentiate both acropetally and basipetally. For about four plastochrones the leaf primordium has no vascular connection with the rest of the plant, emphasizing once more that transport of food materials through meristematic or parenchymatous tissue must be adequate to maintain growth. In grasses and other narrow-leaved monocotyledons several parallel primary veins are present at maturity, and these are again linked by commissural veins running mainly in a transverse direction (Fig. 5.41).

The veins of the leaf are functionally extremely important, since they fulfil the double function of transporting water and dissolved solutes in the transpiration stream and of translocating the products of photosynthesis to other parts of the plant. The complex network of veins must therefore be able to transport materials to and from all the cells of the mesophyll engaged in photosynthesis. It is interesting to note that the distance over which absorption of ions in the leaf is not limited is similar to distances reported to occur between mesophyll and veins.[169] In sugar beet the minor

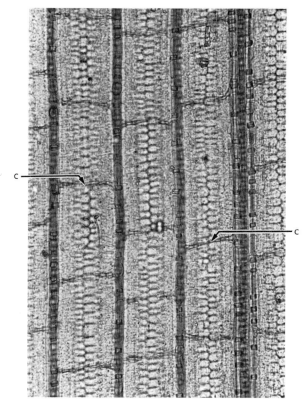

Fig. 5.41 Part of a cleared leaf of *Bambusa*, bamboo, showing the parallel venation. Commissural strands (c) run transversely between the larger veins. × 150.

veins are much more extensive than the major veins and measure 70 cm/cm² of leaf lamina.[202] It was calculated that a minor vein of this species receives translocate from about 29 mesophyll cells; these are situated within an average of 2.2 cell diameters of the vein, or about 73 μm.

The products of photosynthesis must pass from the mesophyll cells to the sieve elements in the minor veins for translocation along the network and into the rest of the plant. Since 1884 it has been thought that certain specialized companion cells were involved in this process. Recently further characterization of such cells has been possible; specialized *'transfer cells'*, believed to function in uptake and export of materials, have been observed to be associated with the minor veins of the leaves or cotyledons of a number of species from at least eight families.[225] As observed with the

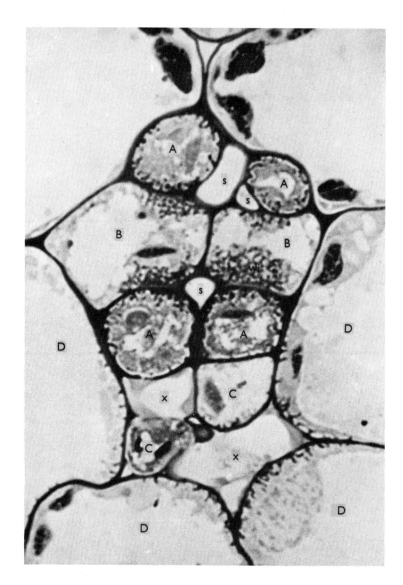

Fig. 5.42 Vascular transfer cells in a minor vein of the leaf of *Anacyclus pyrethrum*. Types A, B, C, and D are all present; note the ingrowths (wi) of the cell wall. s, sieve element. x, xylem.   × 2530. (From Pate and Gunning,[394] Fig. 2, p. 139.)

electron microscope, these cells have unusually dense cytoplasm containing the normal cell organelles, including polyribosomes, chloroplasts with grana and mitochondria with well developed cristae. These cells have protuberances of the wall which project into the cell lumen and result in an increase in the surface/volume ratio; for example, in *Pisum arvense*, the field pea, there is at least a 10-fold increase in area of the plasma membrane as compared with a smooth-walled cell of similar size. Experiments with radio-isotopes indicate that these cells can absorb radioactive compounds both from the transpiration stream and from products of photosynthesis. It is believed, therefore, that one of the functions of these cells may be the export of photosynthates from the leaf mesophyll.

An examination of 975 species showed that among dicotyledons there was a definite association of vein transfer cells with the herbaceous habit. Four different types of transfer cell are now recognized; types A and B are believed to represent modified phloem parenchyma, type C modified xylem parenchyma and D modified parenchyma of the bundle sheath.[394] Examples of all of these types, in which the wall ingrowths occur on different points of the cell wall, are to be seen in Fig. 5.42. A is much commoner than B, but both are common in minor veins compared with the C and D types. Wall ingrowths may occur not only in cells associated with minor veins, but also in those in other parts of the plant (see Chapter 8). Fig. 5.43 shows an electron micrograph of a section through the submerged, dissected leaves of the water buttercup, *Ranunculus fluitans*, in which wall ingrowths can be clearly discerned in the lateral walls of the epidermal cells. Other cells with wall ingrowths are further discussed in Chapter 8.

Transfer cells are believed to be a specific development in which the surface/volume ratio of the protoplast of parenchyma cells is increased by the plasmalemma passing around the wall ingrowths. These cells could then perform more efficiently the function of absorbing solutes from the mesophyll and elements of the vein, passing these to the sieve elements.[394] These cells will no doubt be the subject of extensive further investigation.

The question of what factors control the differentiation of these complex and varied venation systems is clearly not an easy one, and seems to have received little investigation so far. The view that the arrangement of the veins is related to the distribution of growth during leaf development was held by classical botanists such as Goebel and Troll.[191] The present discussion will attempt to put this into rather more specific terms. The evidence for the view[547] that a factor or factors resulting from active meristematic growth may be involved in controlling procambial differentiation has already been discussed in relation to leaf primordia in Chapter 3 (p. 70). On this view, the primary strand, or midrib, of leaves of dicotyledons would differentiate in response to a factor resulting from the apical growth of the primordium. In compound leaves, differentiation of the midrib of

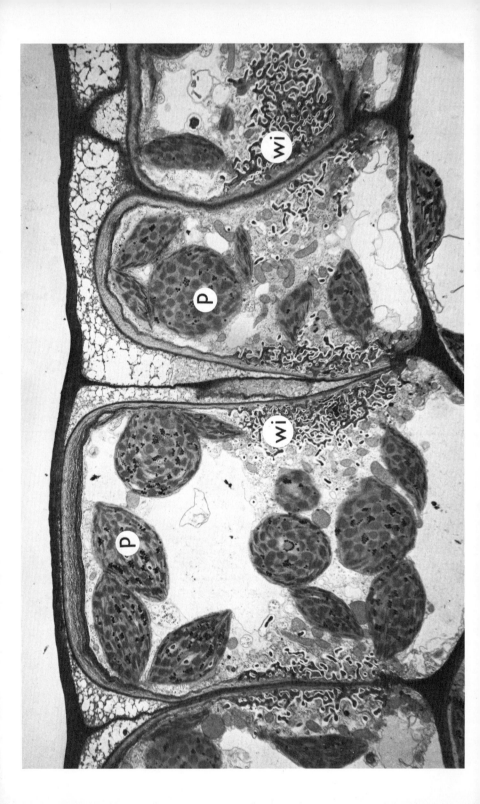

each leaflet could be interpreted in the same way. It is noteworthy that in lobed, serrate or dentate leaves secondary veins are associated with the lobes or teeth of the leaf margin, i.e. with regions of localized marginal growth which may also be producing the postulated factor. Slade[475] has indeed pointed out that the lateral veins of *Acer* differentiate towards and into the crenations, and those of *Prunus* into the serrations of the leaf margin, and has suggested that localized centres of meristematic activity within the leaf may influence the direction of differentiation of the veins. She considers that when intercalary growth is prevalent the factors affecting procambial differentiation may be more uniformly dispersed throughout the leaf blade, leading to the differentiation of the minor veins.

In attempting to understand the factors controlling vein differentiation, the very characteristic venation systems of different species must not be forgotten. Clearly there is overall genetic control of the venation pattern. However, in each species the degree and duration of activity in the various meristems which contribute to the growth of the leaf also vary greatly, as indicated above, and it is not inconceivable that if interactions between these meristems and their products are responsible for the pattern of venation these interactions could be as varied as the systems of venation.

The parallel venation of leaves of monocotyledons is more difficult to interpret in terms of influences from meristematic activity. There is evidence that the vascular bundles are spaced at rather regular intervals. For example, in the leaf blade of *Narcissus* a new bundle differentiates as soon as two existing strands are more than 11 cells apart.[137] Wardlaw[558, 559] has suggested that actively metabolizing meristematic regions may, by some mechanism such as the setting up of tensions, have a direct effect in determining pathways of translocation. In other words, actively growing meristematic regions may both be producing substances necessary for procambial differentiation and also, by acting as a sink and drawing metabolites to themselves, be creating physical tensions which establish pathways along which passage of these substances preferentially occurs. Without attributing to it a similar cause, one can perhaps visualize this system as analogous to the lines of force created by a magnetic field, the meristem representing a pole of the magnet.

In roots and shoots, where apical growth predominates during primary growth, this situation would not be excessively complex (see Chapters 2 and 3). However, in leaves, where, as we have seen, a succession of meristems

---

Fig. 5.43 Electron micrograph of the epidermal cells of a submerged leaf of *Ranunculus fluitans*. Note the wall ingrowths (wi) in the anticlinal cell walls, and the abundant chloroplasts (p).   × 4600. (By courtesy of Dr. B. E. S. Gunning.)

participates in the growth of the primordium, the presumed 'lines of force' would be the resultant of different meristematic influences, would themselves inhibit each other, and might therefore form a much more complex pattern, such as is seen in venation systems. In the leaves of monocotyledons, where the basal intercalary meristem early assumes predominance and marginal growth is less extensive, it might be expected that the main veins would run longitudinally. It is interesting that the secondary veins in leaf primordia of wheat are initiated in the primordium itself and then differentiate both towards the apex and base of the leaf.[465] In the leaf of *Narcissus*, also, the main strands seem to be blocked out as a whole, but subsidiary strands differentiate basipetally,[137] i.e. towards the basal intercalary meristem where the highest rate of cell division is located.

In various experiments, Sachs[441] has shown that xylem strands are inhibited from forming in the vicinity of vascular tissue which is well supplied with auxin. In the case of strands formed approximately simultaneously, he considered that these interactions may not be operative; however, it may perhaps be postulated that the formation of secondary and tertiary commissural veins, which develop later in leaf ontogeny, may take place after the main source of auxin associated with the primary veins is somewhat diminished. Thus they may form in response to 'lines of tension' set up in the mesophyll by the primary strands themselves, which are also no longer capable of inhibiting vascular strands from joining them.

This hypothesis relating to the differentiation of the venation system seems open to experiment. It might be possible, though difficult because of the folding and inrolling of most young developing leaf primordia, both to damage certain sites of meristematic activity and to create artificial sinks. For example, it is known that particular metabolites, such as carbohydrates, will be mobilized towards applied auxin.[53] If it were feasible to carry out such experiments at a sufficiently early stage, it might be possible to interfere with the normal differentiation of the venation pattern.

LEAF STRUCTURE AND ENVIRONMENT

Plants which characteristically grow in certain ecological niches often show a type of structure which is believed to be adapted to that particular environment. This is probably most frequently reflected in the structure of the leaves. Some of the anatomical features of such groups of plants are described below. However, the view that these structural features are adaptations which have evolved to meet the needs imposed by certain habitats, and which confer advantages on plants growing in such habitats, may be questioned in the light of experimental findings, which show, for example, that transpiration is not necessarily reduced in leaves with abundant hairs or sunken stomata. As indicated in the next section, some of these anatomical features may be readily modified and in fact develop in response to particular environmental factors. There may therefore be a causal, rather than adaptive, relationship between environment and structure, though it should be realized that these are not mutually exclusive.

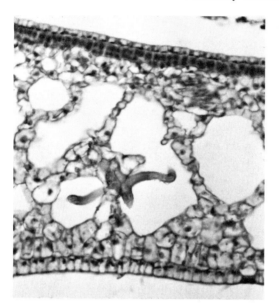

Fig. 5.44 T.S. developing floating leaf of *Nymphaea* sp., water lily. Large air spaces are present in the mesophyll of this aquatic plant. The palisade tissue is just developing. A sclereid is branching into the air spaces.   × 150.

Hydrophytes

The leaves of aquatic plants have certain features in common, though of course their anatomy varies with the species. Submerged leaves are commonly highly dissected, and the leaf is very thin, the mesophyll being reduced to a few layers of cells or none. Stomata may be absent. In floating leaves, the leaf blade is more usually entire, and the leaf is thicker; stomata are often restricted to the upper surface. In leaves of aquatic species the xylem is usually very much reduced, though phloem may be more abundant, and numerous, large intercellular spaces are present between the cells of the mesophyll (Fig. 5.44). In submerged leaves this is usually not differentiated into palisade and spongy tissue.

Some species can grow either in water or on land, and show different leaf forms. This phenomenon is discussed more fully below.

Xerophytes

In species which grow in dry habitats, or in habitats where water is not physiologically available, certain structural features are also common. Leaves of such plants are often rather thick and leathery, with a well developed cuticle and abundant hairs. Well differentiated mesophyll is also present, and there is often more than one layer of palisade tissue (Figs. 5.24 and 5.34). The walls of epidermal and sub-epidermal cells are frequently lignified, and a distinctive layer below the epidermis, the hypodermis, may be present. Leaves of xerophytes have a well developed vascular system and often an abundance of sclerenchyma, either in the form of sclereids or fibres (Figs. 5.34b and 5.45). The leaf is sometimes cylindrical or rolled. This organization is said to protect the stomata, which may also occur in furrows (Figs. 5.45 and 5.46); but in some species of *Carex*, the sedges, the stomata are almost restricted to the abaxial surface and consequently are even more exposed when the leaf is folded upwards.[359]

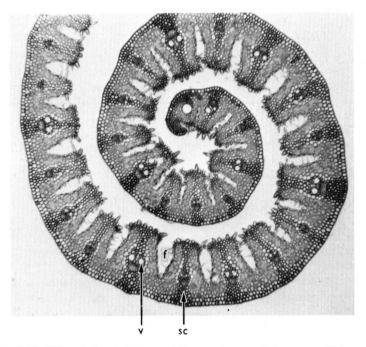

Fig. 5.45   T.S. rolled leaf of *Ammophila arenaria*, a sand-dune grass. This xerophytic leaf has furrows (f) on the adaxial surface in which the stomata occur (see Fig. 5.46). Abundant sclerenchyma (sc) is present surrounding the vascular tissue (v).   ×50.

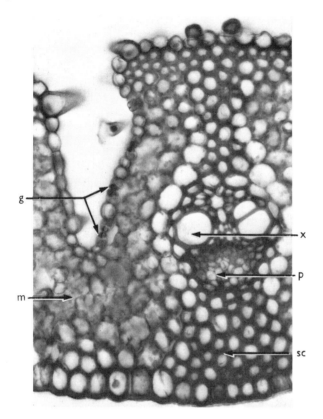

Fig. 5.46 Part of T.S. leaf of *Ammophila arenaria* (see Fig. 5.45). Darkly stained guard cells (g) of the stomata can be seen in the furrows. The mesophyll (m) is relatively undifferentiated. Abundant sclerenchyma (sc) surrounds the vascular bundle, composed of xylem (x) adaxially and phloem (p) abaxially. × 300.

The leaves of some xerophytic plants, for example species of *Sedum* and some halophytes (which grow in conditions where abundant water (salt water) is present but is unavailable because of osmotic conditions), are fleshy, and contain abundant water storage tissue. This usually consists of large thin-walled cells. There is evidence that thickening of the leaf in halophytes is a response to sodium chloride concentration.[9]

Sun and shade leaves

Leaf structure also differs in mesophytes according to the light intensity, resulting in so-called sun and shade leaves in a single species. Sun leaves are usually thicker and more differentiated than shade leaves of the same species, and are more hairy, but have a smaller area of lamina, which may be more deeply lobed. In a study of leaves from different regions of the same tree the leaf structure was found to be considerably modified according to the amount of shading.[585] The mean volume of palisade cells was about 60 per cent less, and that of spongy mesophyll 40 per cent less in the more shaded regions.

In plants of *Impatiens* which were grown in pots and screened so that they had various known percentages of natural daylight the leaves were progressively larger with decreasing amounts of light; the veins were also more widely spaced.[262] In the sun leaves the palisade was well developed and columnar in shape, and there was loosely arranged spongy mesophyll; in leaves of shaded plants the spongy mesophyll was very diffuse (Fig. 5.47). It is interesting that Hughes[262] considers that meristematic activity

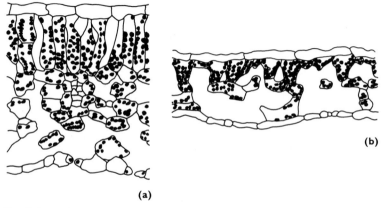

(a)

(b)

Fig. 5.47   T.S. part of the lamina of leaves of *Impatiens parviflora* from plants maintained in (**a**), full daylight, and (**b**), 7 per cent daylight. The shaded leaf is much thinner and the palisade cells are shorter and less highly differentiated. (From Hughes,[262] Fig. 5, p. 165. Published by permission of The Linnean Society of London.)

was similar in the two types of leaves, the much greater expansion of the shade leaf resulting in the differences in structure at maturity. On the other hand, Dostál[150] notes that in branches of the same tree exposed to sun or shade the leaf primordia are already irreversibly determined in the bud. It appears likely that different mechanisms are involved here; in the tree, environmental effects seem to exert an influence at an early stage of leaf development, whereas in an herbaceous plant the effect of shading may be rather on later stages of development, since in this instance leaf primordia will grow only for part of one season, rather than develop over a longer period. Further experiments on this topic are desirable.

CONTROL OF LEAF FORM

This is not the place for a full discussion of variation in leaf form, which has been fully reviewed elsewhere. For example, the reader may refer to papers by Foster[184] on bud scales and foliage leaves, and by Allsopp on heteroblastic development[10, 12] and on land and water forms.[11]

Nevertheless, the phenomenon of **heterophylly** (the occurrence of more than one form of leaf in a single species), and especially what is known of the factors controlling the manifestation of the several forms of leaf, are very relevant to a consideration of leaf development and structure. For the different leaf forms are, in general, a consequence of differential activity of the various meristems involved in leaf development. Heterophylly is

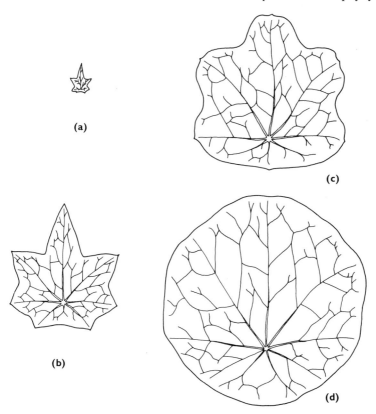

Fig. 5.48 Leaves of *Tropaeolum*. (**a**) Juvenile leaf. (**b**)–(**d**) Leaves of hybrid crosses between *T. majus* and *T. peltophorum* var. *fimbriatum*. The gene U in the absence of L produces roundly lobed leaves. (**b**) Genotype with lu. (**c**) IU. (**d**) LU or Lu. (From Whaley and Whaley,[581] Figs. 1–4, p. 195.)

usually a manifestation of changing conditions, either environmental or during the ontogeny of a plant. It is also discussed in the book by Street and Öpik in this series.[497]

The effects of some other factors on leaf form can also be attributed to their action on certain meristems. For example, in a species cross of *Tropaeolum majus*, leaf shape was found to be controlled principally by two genes which affected differential cell division in early development (Fig. 5.48).[581] On the basis of this work it was concluded that differences in leaf shape were a consequence of differences in growth rates in different dimensions late in development. The pattern was determined during the period of cell division very early in ontogeny although shape differences were not manifest at that stage. This may explain why the leaves of wheat seedlings in which cell division had been prevented by gamma irradiation were nevertheless similar in shape to those of normal seedlings,[230] since it is conceivable that early divisions in the embryo prior to the radiation treatment had already mapped out the leaf shape to some degree, later divisions being less important. That leaf shape depends on the relative activity of the various meristems is also borne out by the curious narrow bladed and 'shoe-string' leaves which develop in tobacco plants infected with tobacco mosaic virus. The 'shoe-string' leaves are radially symmetrical, with no lamina, and result from the inactivity of the marginal meristem.[501]

Effects of environmental factors on leaf shape

Both light and temperature may affect cell division and expansion in developing leaf primordia.[130, 131, 180] High levels of radiation may result in a decrease in leaf area.

Foliage leaves and cataphylls

In many perennial woody plants, the buds are covered by scale leaves or cataphylls during the winter season. Thus the shoot apex produces a sequence of bud scales and foliage leaves, in response to seasonal factors, such as daylength, in the environment. In some species, forms intermediate between foliage leaves and bud scales occur. It was the occurrence of such intermediate forms that led Goebel[216] to consider that bud scales were arrested forms of foliage leaves. As long ago as 1880 he showed[215] that if shoots are defoliated, primordia which would have been bud scales grew out as foliage leaves, or, if defoliation was late in the season, as intermediate forms. Others have performed comparable experiments. As Foster[184] has pointed out, this and other evidence suggests that leaf primordia are initially capable of developing along several pathways but respond to external and internal factors which may affect their development. Subsequently he showed [186, 187] in a series of studies on the hickory, *Carya*, that the divergent development of foliage leaves and cataphylls was established by the time the primordium was 90–190 μm long. Indeed, at 50–90 μm deep cytoplasmic staining and rapid cell division already suggested that the primordium would be a foliage leaf. Subsequently there is pronounced adaxial growth leading to increase in thickness. By contrast, in the potential cataphyll marginal growth has already begun by the time the primordium is 90 μm long. Apical growth ceases early.[186] In primordia destined to develop as intermediate or transitional forms both marginal and adaxial growth occur early, and are somewhat antagonistic.[187] In *Morus* the primordia of bud scales and foliage leaves are not distinguishable with certainty until they are at least 70 μm in length.[117, 118] In *Narcissus* the development of scale and foliage leaves diverges only when the primordium is about 1 mm long, depending on whether cell division is localized at the base of the blade or in the sheath.[136]

It appears, therefore, that the development of a primordium as a cataphyll or as a foliage leaf largely depends on the distribution of activity between the apical, adaxial and marginal meristems; this can apparently be affected both by seasonal factors such as daylength and temperature and by some effect of the older leaves.

Land and water forms: Proserpinaca *and* Ranunculus

As already mentioned, some species exhibit different forms of leaf if submerged or grown either on land or with their leaves above the water surface. Differences between submerged and aerial leaves are often profound; both may occur on the same shoot, apparently as a direct consequence of environmental factors. In *Proserpinaca palustris* McCallum [354] showed in 1902 that if the shoot apices were just below the water level the highly dissected water form leaves developed; if just above the water surface aerial leaves with a toothed margin developed. Later it was shown that formation of dissected leaves could be induced by growing aerial plants in short photoperiods (SD); under long day (LD) conditions the plants formed serrate lanceolate leaves. After treatment with gibberellic acid (GA), plants in SD approached the serrate lanceolate type of leaf.[133, 544] More recently, it has been shown that the other type of leaf can be induced either by submerging shoots or by changing the daylength conditions (Fig. 5.49). In LD and with high light intensity, however, submerged shoots formed expanded leaves.[452]

In *Proserpinaca*, submerged and aerial leaves differ considerably not only in shape (Fig. 5.50) but also in internal structure. Expanded, lanceolate aerial leaves have one layer of palisade and 3–4 layers of spongy mesophyll, and stomata on both surfaces. By contrast, pinnatifid submerged leaves have usually 3 layers of undifferentiated mesophyll and few or no stomata.

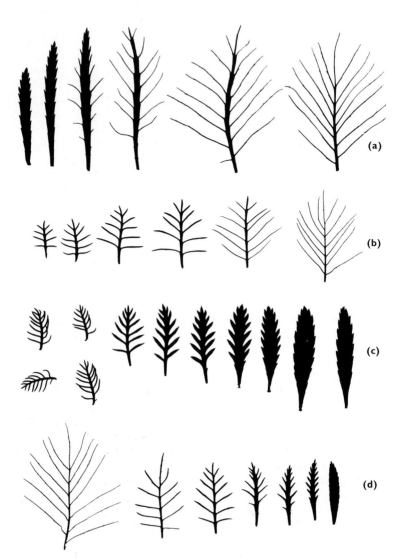

Fig. 5.49 A series of transitional leaves of *Proserpinaca palustris* formed on shoots subjected to abrupt changes of environment. The oldest leaf is at the left of each figure. (**a**) Transition from long day (LD) aerial to short day (SD) submerged leaf form. (**b**) SD aerial to SD submerged. (**c**) SD aerial to LD aerial. (**d**) LD submerged to LD aerial. (From Schmidt and Millington,[452] Fig. 3, p. 268.)

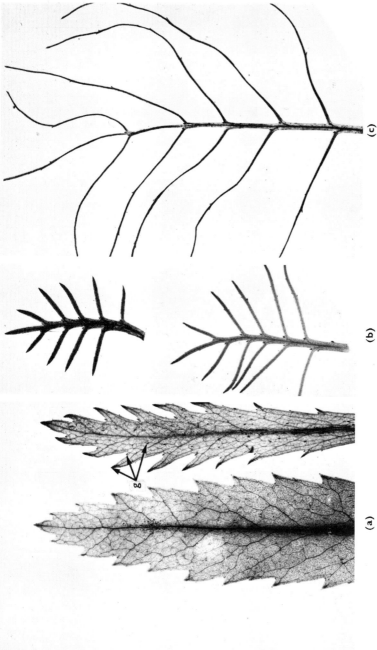

Fig. 5.50 Cleared leaves of *Proserpinaca palustris*, showing variation in leaf form according to the environment. (**a**) Expanded leaves of LD aerial shoots, showing dark glands (g) on the abaxial surface and along the margin. (**b**) SD aerial pinnatifid leaves. There is only a single vein in the midrib and each lobe of the leaf. (**c**) Submerged SD leaf. (From Schmidt and Millington,[452] Figs. 19–21, p. 275.)

(a)

(b)

(c)

Venation patterns also differ. The shoot apex and also early development of the leaf primordia of both types are similar; lobes are formed basipetally at the leaf margin of both types. After the primordium has reached a height of 500–600 μm and has 5 pairs of lobes (in the 5th plastochrone of LD leaves and the 10th or 11th of SD leaves), the development diverges.[452] In this instance, therefore, it appears that the various environmental factors act on the degree of marginal growth in the leaf axis and on the extent of apical and marginal growth of its lobes or pinnae.

In another aquatic species, *Ranunculus flabellaris*, in which the much-dissected submerged leaves differ markedly from the aerial form (Fig. 5.51a

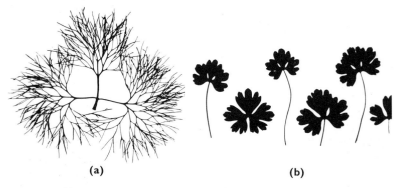

(a)   (b)

Fig. 5.51   Leaf form in *Ranunculus flabellaris*. (a) Dissected leaf from 7th node of a naturally occurring submerged shoot. (b) Expanded lobed leaves from 7th node of naturally occurring terrestrial shoots. (From Bostrack and Millington,[60] Figs. 1 and 2, p. 4.)

and b), low temperature can induce the development of dissected leaves on terrestrial plants. Again the shoot apices of submerged and aerial plants are similar, and environmental factors are considered to act on the rate of formation of lobes in the developing leaves, affecting cell division directly.[60]

The rather puzzling relationship between daylength and submergence, discussed above for *Proserpinaca*, has been investigated by Cook[108] in *Ranunculus aquatilis*. In this species, three forms of leaf occur: submerged, extremely dissected leaves; terrestrial, dissected leaves with much shorter segments, which develop in air under LD; and lobed, more or less entire floating leaves which develop in submerged plants under LD. The dissected and entire kinds of leaf may develop on the same apex in successive plastochrones. Again, the differences depend on the amount and location of activity in the marginal meristems. Various experiments showed that in SD (10 hours of light or less) only dissected leaves of the submerged kind developed, irrespective of whether the plants were grown submerged or on land. In 14-hour photoperiods, only dissected leaves were formed, but these were of the submerged kind in submerged plants and of the terrestrial kind in plants grown in a dry atmosphere. In a 16-hour photoperiod, however, entire leaves were formed on submerged apices, and dissected leaves of the terrestrial kind on apices raised above the surface of the water. Thus in this instance water determines the leaf form only under a regime of long photoperiods, and has no effect in SD. Cook[108] believes that this flexibility in leaf form enables this species to exploit both different and constantly changing habitats. The effects of these various factors on the growth of the marginal meristems have not been studied directly, and leave a promising field for further exploration.

Experiments on Callitriche

Some species of the aquatic genus *Callitriche*, e.g. *C. intermedia*, have linear juvenile leaves, a form which persists in submerged plants. Linear leaves have few or no stomata, elongated epidermal cells and a single vein. Ovate leaves, the land or floating form, have stomata on the adaxial surface, more or less isodiametric epidermal cells and several veins (Figs. 5.52 and

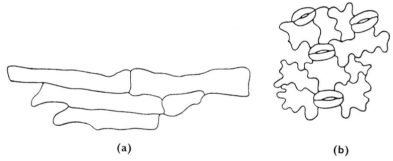

(a)   (b)

Fig. 5.52 Epidermal cells from the adaxial surface of leaves of *Callitriche intermedia*, taken from a midpoint in the leaf at the 11th node. (a) From a linear-leaved shoot. (b) From an ovate-leaved shoot. (From Jones,[289] Fig. 5, p. 100.)

5.53). Again, there is no difference in size or structure of the shoot apex in plants of the two forms.[289] Changes in the form of leaf produced by a shoot can be induced by immersing shoots, or by allowing them to reach the surface of the liquid;[287] similar results were also obtained with *Hippuris*.[356] In an experiment with *Callitriche*, ovate-leaved, floating crowns of three species were transplanted to submerged sites in running water. After 1–2 months the leaves on the shoots of two species were linear with one vein, while those of *C. stagnalis* were considerably narrower but still 3-veined (Fig. 5.53).[287] Young leaf primordia of linear and ovate crowns were similar

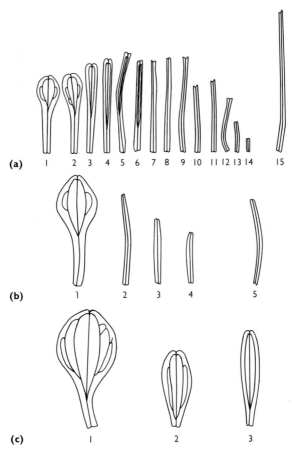

Fig. 5.53 Leaves from ovate-leaved shoots of *Callitriche* spp. which have been experimentally submerged in running water for 1 month or more. (a) *C. intermedia.* 1–14 are leaves from successive nodes after 1 month's submergence, 1 being an original ovate leaf; 15 is a crown leaf from a plant submerged for 3 months. (b) *C. obtusangula.* 1–4 submerged for 1 month; 1 is an original ovate leaf; 2–4 are successive leaves from the crown of an axillary shoot; 5 is a crown leaf from the main axis 3 months after submergence. (c) *C. stagnalis.* 1–3 submerged for 2 months; 1 is an original ovate leaf; 2 a representative leaf of the main axis; 3 the most linear leaf produced on any of the shoots. (From Jones,[287] Text-fig. 1, p. 229.)

in form; the 4th or 5th primordium from the apex showed the first clear difference in form, the presence of 3 veins in primordia which would become ovate leaves, and the assumption of a distinctive shape.

In *C. intermedia* it was possible to bring about the converse change in form and induce ovate leaves from submerged crowns forming linear leaves by submerging them in 30 per cent sea water.[288] These leaves resembled normally occurring ovate leaves in stomatal frequency and epidermal cell shape, but had only one vein. Changes in leaf shape can apparently occur at a fairly late stage of leaf development. It appears from these experiments that immersion or emersion of the crown of the plant affects the degree of marginal growth in developing leaves; further studies of the relative effects of differences in osmotic pressure on cell division and cell enlargement might be instructive. In this connection it is interesting that rosettes of *C. stagnalis* treated with the hormone gibberellic acid (GA) resembled those normally produced from submerged apices.[355] This species does not produce truly linear leaves upon submergence. However, leaves produced after GA treatment showed an increase in the length/breadth ratio of about 50 per cent.

Experiments on Marsilea

The fern *Marsilea*, which has quadrifid adult leaves, has also been the subject of much work on leaf form. Allsopp[5] showed that sporelings grown in aseptic culture on media with 5 per cent glucose produced the land form of leaf, while those on the same medium but with 1 per cent or 2 per cent glucose developed the water form. Land form leaves differ from the water (submerged) form in shape, presence of stomata on the abaxial as well as the adaxial surface, and the shape of the epidermal cells. The water form of leaf could be induced in plants grown in 4 per cent glucose—which would normally have produced the land form—by adding GA to the medium.[6] From these and many other experiments, Allsopp[7] concluded that the internal sugar concentration in the tissues was a key factor, this being affected by other variables.

Using a different species of *Marsilea* and continuous light, Gaudet[200] found that plants in 5 per cent sucrose in liquid medium gave rise to the submerged type of leaf rather than the land form. Etiolation by darkness or far-red light followed by growth in continuous light led to formation of the land form of leaf in the same medium. This occurs both with leaves on the plant and with isolated cultured leaves.[201] Gaudet[199] pointed out that the primordia of both types of leaf are similar until a late stage in development and that the differences are a function of the activity of the marginal meristems. Thus, this activity can apparently be modified by a number of different factors.

Effects of hormonal factors on leaf shape

It has already been mentioned that gibberellic acid (GA) can affect leaf shape. It is known to do so in a number of species, and in different species

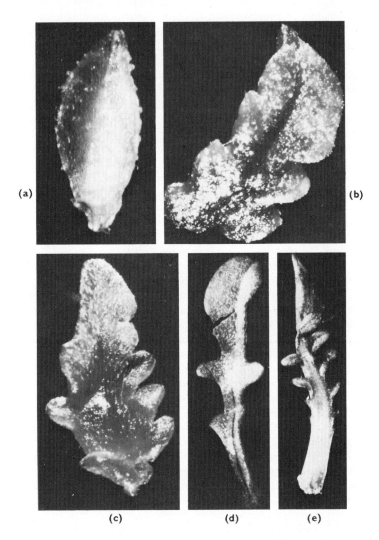

Fig. 5.54 Leaf primordia of *Centaurea solstitialis* excised from the plant and grown in a sterile nutrient medium for $1\frac{1}{2}$–2 months. (a) Leaf which was the 3rd youngest primordium, P_3, at the time of excision, failed to develop any lobes in culture. $\times 50$. (b) P_5, which has developed lobes in culture. $\times 15$. (c) P_6. $\times 12$. (d) P_7. $\times 12$. (e) P_{10}, essentially a miniature version of an adult lobed leaf. $\times 3$. Compare Fig. 5.55. (From Feldman and Cutter,[176] Figs. 11, 14, 15, 17 and 22, p. 43. © 1970 by the University of Chicago. All rights reserved.)

may promote either lobed or entire leaves.[125] It has never been clear, however, whether the GA acts directly on the growing leaf or whether its effects are mediated through the shoot which bears it. Experiments in which both whole plants and isolated leaf primordia were cultured help to answer this question, since it was found that GA affected the leaf primordia even when isolated from the shoot. Recent experiments with a thistle, *Centaurea solstitialis*, showed that if whole plants were grown in sterile culture on media containing GA the leaves produced remained simple and never attained the lobed form of control plants. If plants producing lobed leaves were transferred from control medium to media with GA, subsequent leaves were simple and entire.[175] If individual leaf primordia of *C. solstitialis* were excised and grown in culture, all but the 4 youngest developed as small lobed leaves (Fig. 5.54). If similar primordia were cultured in media with GA, however, the primordia formed fewer lobes and were considerably elongated (Fig. 5.55). The results suggested that GA affected the plane of cell division in the marginal meristem; it may also have affected cell elongation. These effects resulted in elongation of the rachis and the ob-scuring of some lobe primordia.[176]

In some species of *Acacia* the juvenile leaves are bipinnate, compound leaves, whereas the adult foliage organs are entire phyllodes. In *Acacia* a phyllode is equivalent to the petiole and rachis of a pinnate leaf.[47] In a classic study of the development of the phyllode, Boke[47] showed that the distinctive vertical lamina-like phyllode was formed as a result of prolonged activity of the adaxial meristem (Fig. 5.11). Consequently the phyllode is flattened in a plane radial, rather than tangential, to the shoot apex. Activity of the adaxial meristem continues until the primordium is 1000–1500 μm high. Intercalary growth and division of the plate meristem ensue. A comparison with bipinnate leaves and transitional forms emphasized the importance of the relative development of the adaxial meristem and the leaflet primordia, the latter presumably being a consequence of activity of the marginal meristem.

Treatment of phyllode-bearing shoots of *A. melanoxylon* with GA resulted in an increased rate of growth and the development of pinnate juvenile leaves instead of entire phyllodes.[54] It was considered that the formation of pinnate leaves was related to the increased rate of growth of the shoot, but it seems possible that the GA affected the relative amounts of adaxial and marginal growth in the individual leaf primordia. Experiments on culturing excised leaf primordia of *Acacia* might throw light on this interesting problem.

Ontogenetic changes in leaf shape

Many plants show *heteroblastic development*, a change in leaf shape with age. It is often considered that these changes reflect the increasing

nutritional status of the growing plant. The results of a number of experiments are compatible with the view that nutritional factors affecting leaf development are mediated through their effects on the shoot apex, particularly its size.[10, 12] On the other hand, the shape of very young leaf primordia excised and cultured on a sterile nutrient medium is considerably affected by the amount of sucrose in the medium.[499] Young leaf primordia of *Centaurea*, also, are capable of attaining a miniature version of the adult, lobed form when excised before lobe primordia were formed and cultured in isolation (Fig. 5.54).[176] Despite these findings with excised leaf primordia, factors in the whole plant which regulate the rate of leaf development are probably important in controlling its form;[499] their effect must be on the activity and duration of activity of the various meristems in the developing leaf primordium, but we have as yet little or no understanding of the mechanisms involved. There is a need for more precise studies of meristematic activity in leaves.

SENESCENCE AND ABSCISSION

Whatever path of development the leaf has followed, and whatever form it has attained, ultimately it undergoes senescence and abscises, falling from the plant. In deciduous species leaf fall is a seasonal phenomenon; in so-called evergreen species, individual leaves abscise after one or more years of growth and development, but all the leaves do not fall at one time. Senescence is as yet an imperfectly understood process.

The activity of leaves and other determinate organs is regulated not only by changes within themselves but also by interactions with other parts of the plant.[77] The results of one series of experiments in which interactions between parts of bean plants were investigated support the view that senescence-delaying factors (possibly cytokinins), produced in the roots and required in the leaves for protein synthesis, are diverted from the leaves when developing fruits are present, resulting in leaf senescence.[571]

Fig. 5.55 Leaf primordia of *Centaurea solstitialis* excised from the plant and grown for $1\frac{1}{2}$–2 months in a nutrient medium supplemented with the hormone gibberellic acid. The effect of the hormone was to inhibit the number of lobes, and in older primordia to promote elongation of the lobes and of the whole primordium. Compare Fig. 5.54. (a) Primordium aged P_5, the 5th youngest, at the time of excision. No lobes have developed. The primordium is hooked at the tip and has a callus-like development at the base.   × 10. (b) P_7. No lobes have developed other than those already present at the time of excision.   × 12. (c) P_9.   × 10. (d) P_{10}.   × 3. (e) P_{11}.   × 3. (f) Left to right, P_9, P_{10}, P_{11}, and P_{12} at the same magnification. The older primordia have grown to a relatively greater extent than P_9.   × $2\frac{1}{2}$. (From Feldman and Cutter,[176] Figs. 24, 28, 32, 33, 34 and 36, p. 45. © 1970 by the University of Chicago. All rights reserved.)

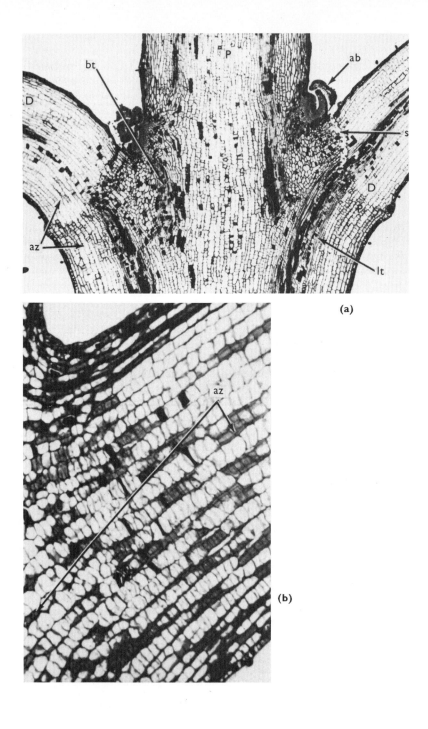

(a)

(b)

However, the mechanisms involved in senescence are still highly contro-versial.

Leaf senescence is usually characterized by a yellowing of the tissues and by various accompanying biochemical changes, such as less efficient or decreasing synthesis of RNA and protein. During the course of senescence various different patterns of metabolism may occur.[467] Structurally, the most important changes related to senescence are those occurring at the time of abscission. This topic is also considered in another volume in this series.[497] Leaf abscission usually follows senescence or injury. Abscission is regulated by various hormones and inhibitors; neither the process itself nor the effects of these substances on structural changes is fully understood. Basically, abscission involves the separation of the leaf from the stem or basal region of the petiole by separation or lysis of cells, and the subsequent formation of a suberized *protective layer* beneath the exposed surface. An *abscission zone* is present at the base of the petiole (Fig. 5.56a and b). It is often externally recognizable by its paler colour; sometimes it is also constricted. Internally, the abscission zone consists of a region of small, thin-walled squat cells with few or no inter-cellular spaces (Fig. 5.56b). Starch is usually absent or in small amount though it may be abundant in adjacent regions. Fibres are either small or absent.[2] The abscission zone is developed to different degrees according to the species. It may be regarded as a region of abrupt structural transition,[575] and is sometimes considered to be a region of weakness.

The *separation layer* typically develops in the distal region of the abscission zone.[2] Starch is often deposited in this region. Substances such as suberin and lignin may be localized in a region several cell layers deep below the separation layer.[576] A protective layer, which minimizes injury to the exposed surface after separation, is thus formed.

In *Phaseolus*, the French or dwarf bean, a considerable increase in the number of tyloses in the tracheary elements takes place in the abscission zone during abscission. This is accompanied by a dissolution of callose in the sieve elements in this region. These changes were accelerated by treat-ment with substances which promote abscission and diminished by treat-ments which inhibit it. It is suggested that the presence of numerous tyloses may cause water stress in the tissues distal to the separation layer, while the dissolution of callose expedites mobilization of materials to more

Fig. 5.56 Abscission regions in cotton. (**a**) L.S. explant showing abscission zones. D, distal; P, proximal; ab, axillary bud; az, abscission zone; bt, branch trace; lt, leaf trace; s, separation.   × 25. (From Bornman, C. H. (1965). Histological and histochemical effects of gibberellins and auxin in abscission. Ph.D. disserta-tion, University of California, Davis, Fig. 16.) (**b**) L.S. cotyledonary node of cotton showing abscission zone (az).   × 75. (From Bornman et al.,[58] Fig. 2, p. 126.)

proximal tissues, both of these effects contributing to localized cellular senescence in the abscission zone.[460] However, Bornman[55] found that the more rapidly abscission was induced in cotton explants, using various regulators of abscission, the fewer tyloses were formed. He concluded that although there was a correlation between abscission and the formation of tyloses it was not a causal relationship.

Separation may occur by dissolution of the middle lamella, by dissolution of both the middle lamella and the primary cell wall, or by mechanical breakage involving non-living elements.[575] Consequently attention has been directed towards changes in the cell wall and in enzymes which dissolve pectin and other cell wall components. In bean, increased pectinase activity preceded separation and the cell wall changes observed by electron microscopy were consistent with pectin dissolution.[369] It was considered, however, that pectin dissolution alone would probably not suffice for separation. In *Gossypium*, cotton, both pectic substances and hemicelluloses are apparently removed from the separation layer during abscission. The removal of pectic substances may be followed by a breakdown in the components of the primary wall, resulting in rupture. Abscisic acid caused a localized lysis of intercellular and cell wall components.[56, 59] The observation of branched plasmodesmata in cells of the abscission zone of flower pedicels led to the suggestion that dissolution between cells might occur at these sites.[275]

It has been known for some years that auxin moving from the leaf lamina delays abscission. (The same is true in developing fruits; see Chapter 7). Explants, which consist of a segment of stem bearing one or more petioles from which the leaf blade has been excised, have been used in much experimental work on abscission, since abscission of the petiole can be accelerated by removing the leaf blade. In fresh explants of cotton, *Gossypium hirsutum*, the presumptive abscission zone does not differ anatomically from adjacent regions of the petiole, but abscission occurs in about 60 hours, following various structural changes.[58] Explants can be readily treated with various hormonal substances and their effects observed. Jacobs[270] has warned, however, that extrapolation of results from explants to the intact plant may not always be valid.

Daylength may affect abscission, long days retarding and short days hastening it. Although most deciduous species are plants of the temperate zone in which leaf fall may be an adaptation to unfavourable winter conditions, some tropical plants also show seasonal leaf fall. It has recently been shown that leaf shedding in the tropical perennial *Plumeria* is a response to short day conditions. Interruptions of the long night prevented the shedding of foliage.[372] This mechanism may afford the plant protection against seasonal drought.

Prior to leaf abscission there is a decrease in the amount of auxin in the leaf and a decrease in the auxin gradient across the abscission zone.[2] (Compare the effect of auxin on the formation of phellogen in the stem, discussed in Chapter 4, p. 116.) As the leaf ages, there are changes in the efficiency and direction of auxin transport in the petiole.[270] Under some circumstances, especially with explants, applied auxin can promote abscission; these effects are interpreted in terms of a stimulation of ethylene production.[195] Much experimental work has been done with ethylene. Recently an abscission-accelerating hormone, now called *abscisic acid* (ABA), has been isolated from young cotton fruits,[391] and other plant parts. This substance is a growth inhibitor which also has effects on dormancy and other aspects of growth and development. In this connection a report of a radiation-induced mutant of *Corchorus olitorius*, in which there was both a retardation of growth-rate and an increase in the rate of abscission, is interesting.[463] A crude water extract of senescent leaves of the mutant accelerated abscission in the parent strain, suggesting the occurrence of an abscission-accelerating substance in leaves of the mutant. Recently it has been reported[312] that part of an extract (believed to be the protein fraction) of the larvae of the boll weevil of cotton accelerated the abscission of cotton flowers and de-bladed petioles. Other proteins, when similarly injected, had no such effect. It is thought that the abscission-promoting substance is probably a protein that is released into the flower bud when the larva moults. The boll weevil normally feeds and deposits its eggs in the flower buds of cotton, and it was the observation that oviposition led to the abscission of the flower buds that prompted this work.

These various regulators of abscission, or the balance between them, clearly bring about the structural changes associated with leaf fall, and some work has been done to elucidate their anatomical effects. By use of explants of the cotyledonary node of cotton plants, it was shown[58] that abscisic acid (ABA), the growth inhibitor phosphon, and GA all accelerated abscission. The auxin indoleacetic acid (IAA) retarded abscission. In all explants, including the controls, some increase in the number of cells in the abscission zone occurred prior to abscission. This was enhanced by GA and inhibited by ABA. The formation of tyloses occurred in all treatments, and was promoted by IAA. These treatments also affected the manner of separation. In the controls rupture of parts of periclinal walls occurred, either along a newly formed wall or through a newly formed cell, disrupting also its contents. With ABA no well defined separation layer was formed and separation took place by lysigenous breakdown of cells; with GA, a separation layer was formed and there was schizogeny at the middle lamella.

If GA was applied to the distal cut surface of the stem, instead of the severed petioles of explants, petiole abscission was accelerated and stem abscission was also induced.[57] Normally no abscission would occur in the stem and there would be no abscission zone. In the stems treated with GA an abscission zone with an active separation layer was formed; separation was similar to that of the petiole. In some species, such as the tumbleweed *Psoralea argophylla*, stem abscission occurs normally,[37] resulting in this instance in the separation of the whole aerial part of the wind-blown tumbleweed. In this instance a separation layer differentiates within the intercalary meristem in one or more internodes near ground level. Cell division precedes abscission of the stem (but not that of the leaves), pectic compounds in the middle lamella region are converted to soluble forms and cell separation takes place. This mechanism seems to bear at-least a superficial resemblance to that induced by GA in cotton explants, and in view of the effects of GA on intercalary meristems (see Chapter 3) and the association of abscission with these meristems in this species, hormonal regulation of abscission in the stem of this tumbleweed might be worth investigating.

A histochemical and autoradiographic study of the distribution of RNA in the cells of bean explants showed that initially there was little difference in this respect between the cells proximal or distal to the region of future separation.[575] After the explants had aged for 24 hours there was an increase in nuclear and nucleolar RNA in cells immediately proximal to the abscission zone. This pattern of localization was accentuated by treatment with ethylene gas, which promotes abscission, for 4 or 8 hours after the period of ageing. There was also a decrease in the amount of protein in cortical cells distal to the abscission zone, and an increase in the cells in and proximal to this region. This increase in RNA and protein after ethylene treatment was very localized, being restricted to 2–6 layers of cells. After ethylene treatment cell division ceased and the cortical cells became disrupted by dissolution of newly formed transverse walls. Ethylene treatment apparently interrupts a sequence of anatomical changes in the explant and accelerates a series of events which result in cell separation. An increase in RNA and decrease in protein in the cells may be a normal preliminary to abscission. It is suggested that applied ethylene becomes effective only after the endogenous level of ethylene in the explants has declined. On this view, the control of cell division in the explant and the series of changes in the cell walls leading to separation would depend on the level of ethylene in the tissues, whether endogenous or applied.

The control of abscission is a complex and rapidly developing field of research, which has currently received impetus in some parts of the world because of the use of defoliants in warfare. The mode of action of each of the hormonal substances involved in abscission has still to be established. When this is known, we shall be in a better position to interpret the structural effects of each substance and the interrelationships between metabolism and anatomy in senescing leaves.

6

The Flower

The flower, which may occur singly or as part of an inflorescence, is formed during the reproductive phase of growth. It develops from a terminal or lateral vegetative shoot apex and results in the culmination of meristematic activity of that particular meristem. Thus the floral apex, like the leaf primordium and unlike the vegetative shoot apex, exhibits determinate growth. In some species, indeed, growth of the whole plant may terminate at flowering, if all the terminal and lateral shoot apices are transformed into flowers or inflorescences. Plants which die after flowering are termed **monocarpic**; it is believed that all their meristems may give rise to flowers.

The classical view of the flower is that it can be regarded as a compressed shoot, in which the sepals, petals, stamens and carpels (and the sterile staminodes and carpellodes, where present) are successive lateral organs. In 1790, in his Theory of the Metamorphosis of Plants, the German botanist Goethe expressed the idea that all the lateral organs of the shoot—leaves, sepals, petals, stamens and carpels—were simply different forms of an idealized lateral member, which he termed a leaf.[16] This view implies that these organs are formed by a single shoot apex. Much later the Belgian botanist Grégoire[218] argued that the floral apex was an organ *sui generis* (unique, not classifiable with others), and did not originate directly from the vegetative shoot apex. The work of Philipson[400] and many others, however, has established that at flowering the vegetative shoot apex, whether terminal or lateral, undergoes various physiological and structural changes and becomes directly transformed into a reproductive apex which will develop into either a single flower or an inflorescence. Flowering is thus merely a stage in the ontogeny of the shoot apex and of the whole plant.

If a single meristematic region, the shoot apical meristem, gives rise to lateral organs as morphologically and functionally diverse as cotyledons, leaves (foliage and scale), bracts, sepals, petals, stamens and carpels, it is reasonable to suppose that during ontogeny it must undergo various rather profound changes. Wardlaw[557] has suggested that during development there is a sequential evocation of genes and an apex passes through different physiological states, which are expressed by the formation of the different lateral organs. Each of these states may be dependent on the preceding one, forming what is known as an epigenetic sequence. A related view is that of Heslop-Harrison,[249] who also considers that each phase of floral development represents the expression of a different gene complement; that is, during ontogeny different floral genes become derepressed sequentially. Again this may be an epigenetic phenomenon, the expression of each set of genes depending on that of the preceding set.

Support for the view that an apex passes through various physiological phases comes from delicate surgical treatments of floral meristems. Cusick[120] bisected young floral meristems of *Primula bulleyana* in various stages of development. New floral apices were formed from the original half apices, and if the bisection was carried out some time prior to sepal formation these apices gave rise to complete flowers. If the cuts were not made till a later stage of development, the new flowers lacked certain organs along the edge adjacent to the wound. These results are considered to support the view that a floral apex passes through a succession of physiological states which regulate the formation of each kind of organ in turn. Similar results have recently been obtained with developing flowers of *Portulaca*.[486] Work in which young floral buds of *Aquilegia* were cultured on sterile nutrient media also indicates that each set of floral organs has rather specific requirements for growth.[502, 503] Analysis of the protein complement of various vegetative and floral organs of the tulip by acrylamide gel electrophoresis showed chemical differences between the various lateral organs, and also between vegetative apices and those induced to flower.[30] Indications are, therefore, that the various lateral primordia of the shoot apex during its progression from the vegetative to the reproductive phase are a consequence of biochemical changes in the apex, and themselves have different growth requirements and a different chemical constitution—just as they also have structural differences, described below.

The change from the vegetative to the floral phase of development, the most profound ontogenetic change in plants, is brought about by various environmental factors, which are still not well understood. Since the reproductive apex is a transformation of the vegetative apex, the changes are first manifest in the apical region and much work has been devoted to correlating structural changes in the apex with various physiological treatments which induce flowering. Even in 1790—perhaps before—it was understood that the flowering phase could be accelerated or delayed.[16] By the use of modern techniques of staining, histochemistry and autoradiography, changes can be detected in the apical meristem within 16 hours after treatment to induce flowering.

Various factors are now known to induce flowering. These are described in numerous books[171, 257, 443] and articles and will be mentioned only briefly here. They are also discussed in another book in this series.[497] Many plants respond to *photoperiod*, i.e. the relative lengths of light and darkness within a 24-hour period. Some plants *(short day plants, SDP)* flower in short days, i.e. with a long, unbroken night, others *(long day plants, LDP)* in long days, i.e. with a short night, and still others are *day neutral*. Some plants require several LD or SD in order to flower, others only one. Each appropriate photoperiod is known as an *inductive* day or night. Photoperiodic treatments are supposed to induce the formation in the leaves of a hormone, *florigen*, which is transported to the apex; but this hormone has never been identified. Temperature may also be important, especially a period of cold treatment *(vernalization)*, and in some rosette species the hormone gibberellic acid (GA) can induce flowering. Various combinations of these factors may be required.

In order to flower, most plants must have attained a certain age or stage of development, a condition known as 'ripeness to flower'. In some species, the position of the shoot with respect to gravity is important, and bending the shoot into a more or less horizontal position promotes flowering.[334] This is true in various commercial crops, such as apples and cherries,[523, 569] and pineapple.[535] Flowering in pineapple is also stimulated by auxin, perhaps as a consequence of ethylene production, and also by acetylene.[205] There is a tremendous bulk of literature on the physiology of flowering. For reasons of convenience, physiologists have tended to work with certain species, notably those which require a single inductive night. Some case histories of the plants most commonly studied have recently been published.[171]

Attempts have now been made to investigate the factors promoting flowering in isolated apices or other tissues maintained in sterile culture. For example, flowering can be induced in cultured vegetative apices of species of *Carex* if a leaf 5·7 cm long is included in the explant. Stimuli produced by growing roots were found to be essential for normal branching of the inflorescence.[478] Segments of internode from the inflorescence of tobacco, variety 'Wisconsin 38', formed callus in culture, and the callus gave rise to floral buds. Without sugar in the medium the primordia were vegetative.[3] In the SDP *Plumbago indica*, flowers could be obtained from internodal segments of vegetative plants kept in LD, if the cultured segments were themselves kept in SD.[379, 380] Neither roots nor expanded leaves were essential for floral induction. In experiments with *Cichorium*, it was shown that even buds formed in cultured segments of root could be induced to develop as inflorescences.[395] The aim of experiments such as these is to establish the metabolites and other conditions necessary for the induction and development of floral meristems.

Flowers or inflorescences usually develop from the main shoot apex or from lateral buds, and are thus typically terminal or axillary in position. In a few species flowers may occupy leaf sites in the phyllotactic spiral, or other extra-axillary positions on the shoot apex.[122,124,419] In such species not only does the main shoot apex remain vegetative, but the floral meristems have no evident preliminary vegetative phase; that is, the meristem develops directly into a flower. In some woody species, e.g. *Theobroma cacao*, the source of cocoa, flowers are borne on the thick, woody stem itself; this phenomenon is known as **cauliflory**. In some instances such flowers originate from axillary buds which have become engulfed by the stem tissues during secondary growth.

Classical morphologists have divided inflorescences into various types—racemes, cymes, umbels, capitula, and so on. These various structures, however, may not really be as distinctive as they appear; their differences may be attributable to differential growth. As Philipson[402] has pointed out, the difference between a cyme and a raceme is one of degree, and depends on the relative amounts of growth of the terminal and lateral buds. Even analysis of the highly complex inflorescences of some palms sometimes reveals that they are constituted by repetitions of quite simple branching systems.[512]

Whether a flower is **hypogynous** (with a superior ovary, i.e. with the floral parts inserted below the ovary), **epigynous** (with the floral parts above the ovary), or **perigynous** (floral parts surrounding the ovary) may also be a consequence of differential growth of the receptacle and floral organs during development. Moreover, flowers of each type, but of different species, do not always form in the same way. Thus Kaplan[300] has pointed out that in the epigynous flowers of *Pereskia* (Cactaceae) the formation of a floral cup, resulting in the other floral parts lying above the ovary, takes place only after the formation of the floral organs; whereas in epigynous flowers of *Downingia* (Campanulaceae) vertical growth of the periphery of the floral meristem exceeds that at the centre in an early stage of development, at the time of sepal formation. The differences between these types of inferior ovary thus reflect a difference in the time that the floral bud becomes concave during organogenesis. In the perigynous flowers of *Portulaca* (Portulacaceae) growth occurs in a plane perpendicular to the axis, resulting in formation of a cup-shaped structure with the carpels attached at the base. Thus perigyny develops differently from the more familiar type found in the Rosaceae.[485]

Flowers may also be radially symmetrical *(actinomorphic)*, e.g.

buttercup (*Ranunculus*), or bilaterally symmetrical *(zygomorphic)*, e.g. sweet pea (*Lathyrus*). Where the plane of symmetry of zygomorphic flowers is radial to the main shoot axis, it is easy to envisage a physiological asymmetry resulting from the presence of a bract on one side subtending the flower, and the main apex or axis on the other; but asymmetry in the tangential plane is less easy to interpret.

Flowers are sometimes said to be perfect, if they possess both stamens and carpels, or imperfect, if they lack either of these organs. Unisexual flowers having stamens but no carpels are called staminate, those having carpels but not stamens are called carpellate or pistillate.

Elongation of the flower stalk

After determination of the floral meristem, the flower or inflorescence must be raised up if pollination is to be effected. This is especially well seen in many aquatic plants with submerged shoots. For example, the peduncles of water lily flowers may elongate to a length of several feet. The intercalary meristem which effects this prodigious growth is shown in Fig. 3.10.

Söding[484] found that whereas growth of the flower stalk of several species was reduced by removal of the flower or inflorescence, it could be increased again to some degree by replacing the flowers on the stump with a gelatine seal between the cut surfaces. More recently, Sachs[436] has investigated the control of intercalary growth in the **scape** of *Gerbera* (a scape is a floral stalk devoid of leaves). He showed that in this species elongation is a result of the activity of an **intercalary meristem** below the receptacle. Both cell division and cell elongation contribute to the growth of the scape, but these types of growth are clearly separated. Sachs noted that growth of the scape ceased if it was decapitated, i.e. if the young inflorescence was removed. By removing the flowers only, he showed that the receptacle and bracts apparently contributed to the normal growth of the intercalary meristem by affecting cell division, since this did not decline as rapidly as in decapitated scapes. Applied GA and IAA enhanced the growth of decapitated scapes, but even when combined did not maintain meristematic activity at the level of that in intact scapes. The normal control of scape elongation by the inflorescence probably involves the synthesis and transport of at least two substances; a sequence of growth substances may perhaps be required at successive stages of cell maturation. Other work on growth of flower stalks is reviewed by Sachs.[436]

TRANSITION TO FLOWERING

Much work on flowering has centred on the earliest stages. In particular, anatomists have recently concerned themselves with the early changes which occur in the apex as a result of inductive treatments; at this stage the apex is said to be undergoing the **transition to flowering**. Essential descriptive anatomical studies form the basis for more recent work in which attempts have been made to investigate chemical and structural changes at the cellular level, by techniques such as histochemistry, autoradiography and electron microscopy. In the most useful and precise work inductive treatments are given and the apices killed and fixed at a known time after the beginning of the inductive period. Species which form a certain, set number of leaves before flowering, and others induced to flower by photoperiod or treatment with acetylene, have all been used in these studies.

Anatomical changes

Flowering is often preceded by an elongation of internodes, and one of the earliest signs of the transition to flowering is the precocious development of axillary buds. In the SD dicotyledon, *Chenopodium album*, these rapidly developing axillary buds were observed after 4 SD.[209] In the grass *Lolium temulentum*, which can be induced to flower by only one LD, axillary bud sites actively incorporated precursors of nucleic acids and this activation of the bud sites was one of the earliest changes resulting from floral induction.[314]

At the transition to flowering, the apex usually changes markedly in shape and sometimes in size. These changes occur both in the formation of a single flower and of an inflorescence (Figs. 6.1–6.3). Activity of the rib meristem increases.[500] Often the apex becomes more elongated and conical, e.g. *Liriodendron*. In sedges it may enlarge in both the vertical and horizontal planes.[476] In *Chrysanthemum* the apex increases in area nearly 400-fold within a few hours, during the formation of the capitulum, a truly striking change.[459]

Linked with and preceding this change in shape is a rise in **mitotic index** in the apex. The mitotic index is the percentage of nuclei that are engaged in mitosis at the time of fixing. For example, in *Xanthium*, one of the species traditionally used in studies of floral induction, a significant increase in mitotic index in the apex was observed within 24 hours, and on one occasion 16 hours, after the beginning of the inductive treatment.[507] In *Pharbitis nil*, another species well known in studies of flowering, an increase in mitosis was noted on the second day after induction.[42] Increased mitotic activity distal to the rib meristem and below the central zone was also the first change observed in apices of several other species of both LDP and SDP, following induction.[579] Other changes at the cellular level include a change in the amount and distribution of endoplasmic reticulum (ER), first visible in *Chenopodium* 3 hours after the end of the first long night. There is also an increase in the number of dictyosomes per cell, and perhaps of the enzyme acid phosphatase. These changes may be a

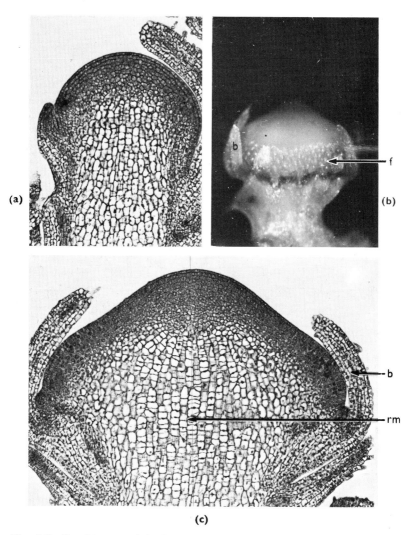

Fig. 6.1 Development of the inflorescence of marigold, *Calendula officinalis*. (a) L.S. very early transition apex, dome-shaped and with a two-layered tunica. ×100. (b) Side view of a living apex which has formed the involucral bracts (b), some of which have been removed, and several ranks of florets (f).   ×30. (c) L.S. young inflorescence apex, which has now broadened considerably (compare (a)). Bracts (b) have been formed and the first floret primordia are just forming; the rib meristem (rm) is conspicuous.   ×100.

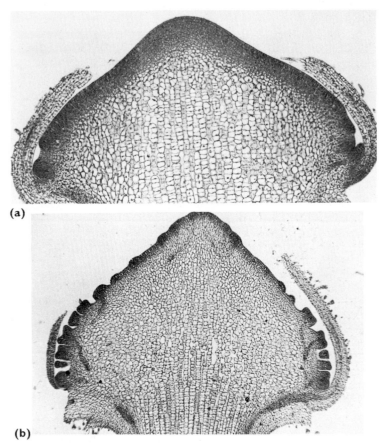

(a)

(b)

Fig. 6.2 Inflorescence apices of *Calendula*. (**a**) The apex is now still broader than in Fig. 6.1c, and has formed several floret primordia at the periphery.   × 75. (**b**) Floret primordia have now been formed over much of the surface of the apex. Compare Fig. 6.3a.   × 38.

result of the inductive treatment itself, or a consequence of the increased rate of mitosis following induction.[207]

During the transition to flowering the typical tunica-corpus organization of the vegetative apex (see Chapter 3) may become modified, as also does cytohistological zonation, if recognized. Changes differ somewhat from species to species, but are fairly similar whether an inflorescence or a single flower is formed. Gradually the cells in the central core of the apex become highly vacuolate, in contrast to the smaller-celled, densely staining layers which form an outer covering or mantle. This structure is termed a

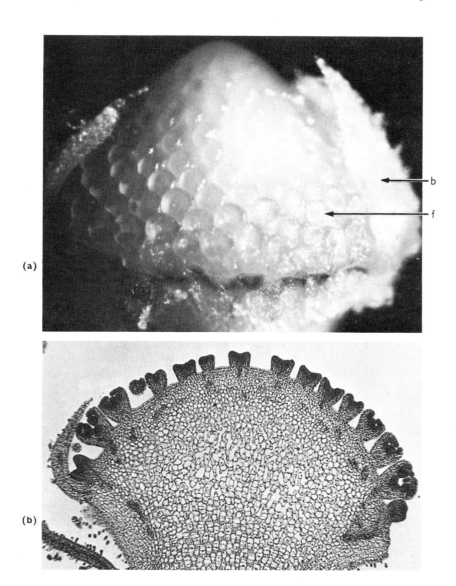

(a)

(b)

Fig. 6.3 Inflorescence apices of *Calendula*. (a) Side view of an apex which has formed bracts (b) and spirally arranged floret primordia (f) almost to the summit. Compare Fig. 6.2b. × 60. (b) L.S. inflorescence apex which has formed floret primordia over the whole area. × 50.

mantle-core type of organization. The term mantle comes from the French *manchon méristèmatique*, a term used by Grégoire. In the capitulum of the common daisy, *Bellis perennis*, the *manchon* is said to be formed by expansion of the peripheral meristem over the surface of the apex.[400]

As an example of an inflorescence we may take the capitulum of the common garden marigold, *Calendula officinalis*. Before flowering begins and in early transition, the apex is domed and has a two-layered tunica (Fig. 6.1a). The apex then grows both vertically and horizontally and forms a broadly-based cone (Fig. 6.1b, c); the active rib meristem is clearly discernible. Primordia of the involucral bracts are being formed, but most of the apical surface is still naked. The apex no longer has the organization of a vegetative apex, but neither does it have an evident mantle–core structure. After further growth the proportional size of the central parenchymatous core increases, and floret primordia are formed at the periphery of the inflorescence apex (Fig. 6.2). An external view of an apex at a similar stage is shown in Fig. 6.3a. Gradually floret primordia are formed in acropetal sequence (Figs. 6.2 and 6.4), until they occur over the whole surface of the meristem (Fig. 6.3b). During this time the apex has continued to enlarge (note the magnifications of the Figures). In Fig. 6.4 scanning electron micrographs of the same inflorescence apex in side view and as seen from above are shown; the contact parastichies (see Chapter 3, p. 46) in which the floret primordia lie are clearly evident, 21 in one direction and 34 in the other. The system of spiral phyllotaxis is thus $(21 + 34)$ on this apex. As in other Compositae[401, 403] the involucral bracts do not subtend florets, and the florets on the surface of the apex are not associated with bracts. In the elongated capitulum of *Dipsacus* (Dipsacaceae), the teasel, however, the involucral bracts do subtend flowers (Fig. 6.5).

One of the plants in which the induction of flowering has been most exhaustively studied is *Xanthium*, the cocklebur. *Xanthium* is monoecious; the terminal inflorescence is staminate, whereas those developing from axillary buds are pistillate.[444] Stages in the development of the inflorescence from the vegetative apex, as seen in dissected apices, are shown in Fig. 6.6. Although the reproductive apex does increase in size, it does not become as conical as that of *Calendula*. The vegetative apex of *Xanthium* has a single-layered tunica. The outermost layer of the corpus is distinctive, and consists of rather large cells (Fig. 6.7a). There is a peripheral zone of small cells, and an evident rib meristem.[204, 579] Apices of plants given one in-

---

Fig. 6.4 Inflorescence apex of *Calendula* viewed with the scanning electron microscope. (a) Looking down at a slight angle on the surface of the apex; the component cells can be discerned. Contact parastichies, 21 in one direction and 34 in the other, are evident. The florets are thus regularly arranged in spiral sequence on the receptacle. (b) Side view. b, bract; f, floret.   × 50. (Photographs by Dr. R. H. Falk.)

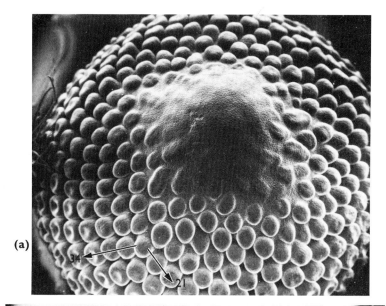

(a)

(b)

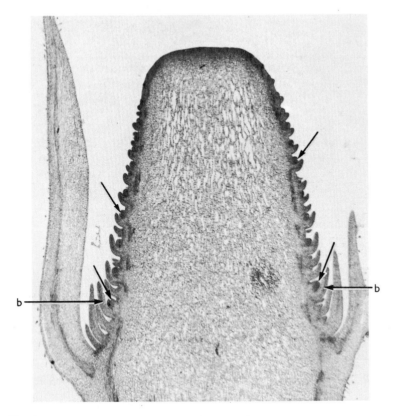

Fig. 6.5 L.S. inflorescence apex of *Dipsacus sylvestris*, with bracts (b) subtending floret primordia (arrowed).   × 28.

ductive night show a change in form two days after the end of the long night.
Considerable mitotic activity occurs in cells between the central zone and
the rib meristem, and an increase in ribonucleic acid (RNA) is also evident
(Fig. 6.7a, b). Seven days after induction the mantle-core type of zonation
is established.[204] Floret primordia are being formed 10 days after induction
(Fig. 6.7c). If plants are given more than one inductive photoperiod, morphological changes occur more rapidly; for example, 8 days after two
inductive photoperiods apices had well developed flower primordia subtended by bracts (Fig. 6.7d). Fig. 6.7a–d may be compared with the external
views in Fig. 6.6.

Somewhat similar changes occur during the development of a single
flower. In the floral apex of *Nuphar lutea*, for example, a mantle-core type

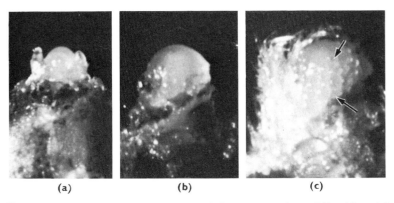

(a)   (b)   (c)

Fig. 6.6 Side views of living developing inflorescence apices of *Xanthium*. (a) Transitional apex. Compare Fig. 6.7b. (b) Apex forming the first floret primordia. Bracts have been removed. Compare Fig. 6.7c. (c) Apex with several ranks of floret primordia (for example, running vertically between the arrows). Most of the bracts have been removed. All × 45.

of organization gradually becomes established, and the apex gives rise to a sequence of floral organs instead of bracts and florets (Fig. 6.8).

The rather complicated grass inflorescence is shown in Fig. 6.9, which may be compared with the vegetative stages shown in Fig. 5.4. Fig. 6.9a shows an early stage of inflorescence development, in which 'double ridges' can be seen. Later the primordia of the individual spikelets grow out (Fig. 6.9b), and later still the primordia of the glumes and lemmas of the grass flower develop on the spikelets (Fig. 6.9c). During this process a considerable increase in size has taken place.

Méristème d'attente

The view that a central, inactive region, the ***méristème d'attente***, is present in vegetative apices, and becomes active at the time of flowering, has already been discussed in Chapter 3.

This view is based not only on the distribution of mitoses within the apical region, but also on the cytological characteristics of the cells. For example, at the onset of flowering in *Lupinus* the cells in the tunica and at the summit of the corpus become more meristematic in appearance, containing numerous small vacuoles and a rather homogeneous chondriome.[74] In the apex of *Aster sinensis*, of which an ontogenetic study was made from the embryo up to the time of flowering, cytological differences became established after the formation of the third leaf primordium. At this time the cells at the summit of the tunica and corpus had larger vacuoles, smaller nucleoli and a slightly different chondriome from the laterally situated cells forming the **anneau initial**. In the inflorescence apex, however, the cells were more uniform, with large nuclei and nucleoli and numerous small plastids.[322]

Much of the work on the distribution of mitoses in the apex, to be discussed below, has stemmed from this theory of apical organization, which is more fully reviewed elsewhere.[125, 203, 385]

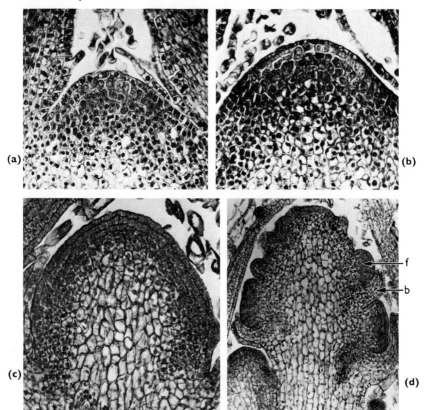

Fig. 6.7 L.S. developing inflorescence apices of *Xanthium*. (a) Before the beginning of floral induction, stained with pyronin.   × 230. (b) 2 days after 1 long inductive night, stained with pyronin. The darker staining indicates an increase in the amount of RNA.   × 230. (c) 10 days after 1 inductive photoperiod. The apex has increased considerably in height.   × 160. (d) 8 days after 2 inductive photoperiods. Floret primordia (f) are present in the axils of bracts (b).   × 120. (From Gifford,[204] Figs. 11 and 13, p. 132, and Wetmore, Gifford and Green,[579] Figs. 10 and 14, p. 264. AAAS Pub. No. 55 ; copyright 1959 by the American Association for the Advancement of Science.)

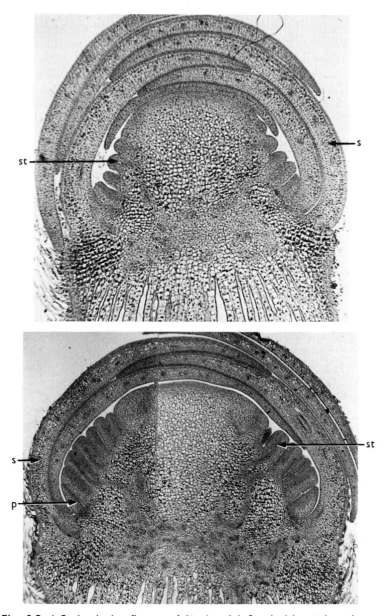

Fig. 6.8 L.S. developing flowers of *Nuphar*. (a) Sepals (s), petals and some stamen (st) primordia have been formed. × 50. (b) Sepal (s), petal (p) and stamen (st) primordia have been formed; the floral meristem is just beginning to form carpel primordia. × 35.

Intermediate apex

As mentioned in Chapter 3, a condition of the apex considered to be intermediate between vegetative and reproductive apices has recently been observed in several species. This state occurs in plants that have been maintained for long periods in non-inductive conditions. Under these conditions the zonation characteristic of the vegetative apex may disappear, and activation of the central zone, or *méristème d'attente*, may begin.[39] In apices of plants of the SDP *Amaranthus retroflexus* maintained for 60 days under 16-hour photoperiods the central axial cells were small and meristematic, had very small vacuoles and many organelles. Nucleoli were intermediate in size between those in vegetative or reproductive apices of plants maintained in SD.[388] Such apices thus have some of the characteristics of vegetative meristems—vacuolate cells, little RNA in the cytoplasm of the tunica cells, small nucleolar volume in the corpus cells, and the continued production of foliage leaves—and some of the features of reproductive apices—greater activity and more RNA in the central axial cells, and increased nucleolar volume in the tunica cells (compare Fig. 6.10, a–d). In *A. retroflexus* flowering is eventually attained even under LD conditions, but this is not so in all species. Because these apices seem to represent something of a compromise between vegetative and reproductive apices of the same species, they have been termed 'intermediate'. Although perhaps merely a matter of semantics, one may question whether this term is entirely appropriate; for the principal criterion of a vegetative apex is the production of leaf primordia, and this continues apparently unchanged in intermediate apices. However, recognition of such a phase in the ontogeny of the apex, whatever it is called, is clearly very important. In some species, plants in this phase of development have acquired great sensitivity to inductive conditions, and thus may truly be close to flowering; in other species, intermediate apices may actually flower later than axillary buds.[39] In *Cannabis sativa*, hemp, the intermediate stage can be considered as a very slow transition to flowering, which in some genotypes is never completed under non-inductive conditions.[254]   Bernier[39] suggests that most of the characteristics of the intermediate stage, which he prefers to regard as a 'waiting stage' rather than a preparation for flowering, may be attributable to an increase of gibberellin in the apical tissues. This suggestion no doubt stems from the observation that in some species treated with gibberellin growth of only the central zone in the apex is stimulated. Since apices change with time even in non-inductive conditions, it is important

Fig. 6.9 Inflorescence apices of *Triticum aestivum*. (a) Double-ridge stage. (b) Spikelet primordia growing out. (c) Glume and lemma primordia present on the spikelet primordia. The panicle is still very compressed and compact. Compare Fig. 5.4.   × 50. (By courtesy of Dr. B. C. Sharman.)

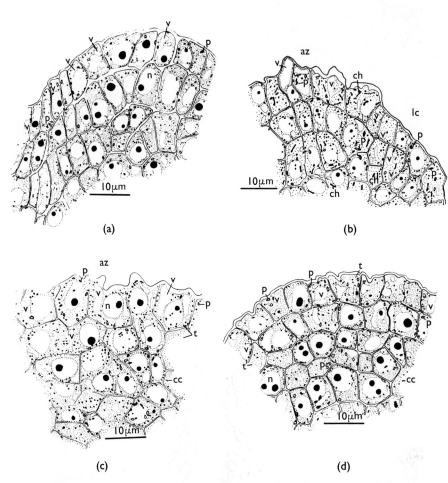

Fig. 6.10 Cytological details of apices of *Amaranthus retroflexus*. (a) and (b) from plants grown in an 8-hour photoperiod. (a) Transition apex. All cells appear meristematic, with small vacuoles (v). (b) Apex becoming reproductive. Vacuoles (v) have enlarged, nucleoli (black) have decreased in size. (c) and (d) from plants grown in a 16-hour photoperiod, which have formed 16 leaves. (c) Intermediate apex. Cells of the flank meristem and the axial central cells have small vacuoles (v); tunica cells (t) have large vacuoles, but also large nucleoli (shown black; nuclei (n) outlined). (d) Reproductive apex. Tunica cells are quite vacuolate and plastids (p) are present; the central cells appear meristematic. az, axial zone; cc, central cells; ch, chondriocont (other organelles stained by the technique used); lc, lateral cells; n, nucleus. (After Nougarède, Gifford and Rondet,[388] Figs. 18, 19, 20 and 21, pp. 288 and 289.)

to note, as Nougarède[385] has cogently pointed out, that the transition to flowering can occur in different ways according to the environmental conditions to which the plants have previously been subjected.

Histochemistry and autoradiography

Where considerable structural changes are taking place, they must certainly be accompanied, indeed preceded, by biochemical changes, since the structure of an organ may be held to reflect the physiology of its growth. Accordingly, various techniques have been applied to apices which have been induced to flower, in an effort to detect early chemical changes in the tissues. Material has been stained with dyes specific for certain substances, usually with an appropriate control from which the substance under study has first been extracted, and apices have been supplied with radioactively labelled precursors of various substances and the technique of autoradiography, explained in Chapter 2, applied. These techniques have yielded some extremely valuable results, but here, as in other aspects of anatomy, there is a great need for the development of refined, delicate biochemical and physiological techniques.

Because physiologists, using analogues of some of the precursors of the nucleic acids, have shown that in different species synthesis of DNA or of RNA is essential for flowering (see Cutter[125]), much histochemical work has been concerned with changes in the amount and distribution of nucleic acids in the apex. An early sign of floral induction is the change in distribution of cells showing an affinity for the dye pyronin; sometimes this affinity is termed pyroninophilia. This dye stains RNA; its specificity can be tested by extracting RNA from some sections with the enzyme ribonuclease. Wherever stain is present in the test sections, but not in the extracted ones, RNA can be said to be present. In vegetative apices, cells with dense cytoplasmic RNA are usually restricted to the flanks; but a few days after floral induction the concentration of RNA, as revealed by affinity for the dye, tends to increase and to be more uniformly distributed throughout the apex.[210, 322] In the SDP *Chenopodium album*, the concentration of RNA increased after 2 SD, and again after 4 SD (Fig. 6.11a, b).[209] In apices of *Pharbitis* and *Xanthium* an increase in RNA is observable within one day after the long inductive night. In *Xanthium* this increase is not observed if the long night is interrupted with 10 minutes of light, indicating that it is a true photoperiodic phenomenon.[204]

In *Datura stramonium*, however, which is not photoperiodically sensitive but forms a terminal flower after the 7th or 8th leaf primordium, although there was an increase in RNA at the time of flowering the axial cells had a lower concentration of cytoplasmic RNA than the peripheral cells in both vegetative and transition apices.[112] In the grass *Lolium temulentum*, RNA was uniformly distributed in the vegetative apex, but there was a localized

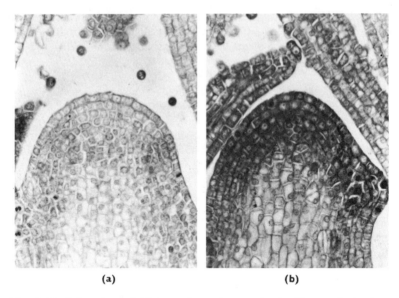

(a)   (b)

Fig. 6.11 L.S. apices of *Chenopodium album*, stained with pyronin to show RNA. (a) Vegetative shoot apex; staining is light and almost uniform. (b) Inflorescence apex after 4 inductive cycles, showing an increase in RNA, especially in the corpus region. Both ×300. (From Gifford,[204] Figs. 1 and 2, p. 129.)

increase at spikelet sites by the 3rd day after induction.[313] In pineapple apices treated with acetylene, the cells of the apical zone showed an increase in cytoplasmic RNA 6 days after treatment.[205] This change in the amount and distribution of RNA following floral induction thus seems to be a rather general phenomenon. As Gifford and Tepper[210] have pointed out, these observations are an interesting and important supplement to the work demonstrating that in some species RNA synthesis is a necessary preliminary to flowering.

By extracting and analysing proteins by acrylamide gel electrophoresis, Barber and Steward[30] have demonstrated changes in the content of soluble proteins in the shoot apex following floral induction. This technique, useful as it is, reveals little concerning the distribution of these substances in the apical tissues. However, the distribution of total protein and of histones in vegetative and induced apices has also been studied by using staining techniques specific for proteins or for basic proteins. In *Chenopodium*, the concentration of total protein was much greater after 4 SD than in vegetative apices. On the other hand, there was a marked decrease in staining for histones after 5 SD.[210] A comparable decrease in nuclear histone was not observed, however, either in *Datura*[112] or *Lolium*.[313]

Both the *méristème d'attente* theory of apical organization, and experimental work indicating that DNA synthesis is essential for flowering, have led to considerable work on the amount and distribution of DNA synthesis and of mitosis in vegetative, transitional and floral apices. Many workers have shown that at the transition to the flowering phase mitosis in the

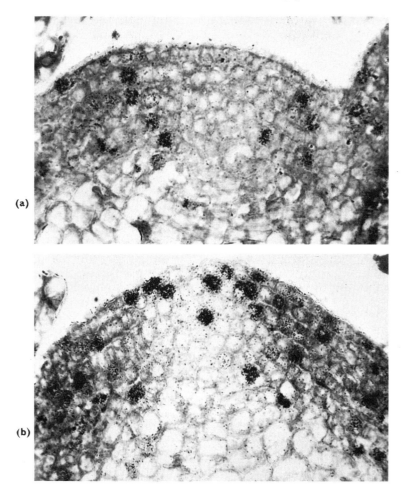

Fig. 6.12 Autoradiographs of the shoot apex of *Sinapis alba*. Both apices received 3 μc of ³H-thymidine for 4 hours. (a) Vegetative apex of a 6-day-old plant grown in long days (16 hours light) since sowing. (b) Transitional apex, from a 9-day-old plant grown in long days since sowing. More nuclei in the central region are labelled. Both × 680. (From Bernier,[38a] Figs. 51 and 53.)

central axial region of the apex is enhanced over that occurring in the same region of vegetative apices.[39, 74, 322] Usually this is based on autoradiographs which show more labelled nuclei in transitional and floral apices (Fig. 6.12). However, the rate of mitosis and the mitotic index should also be taken into account. Corson[111] has recently analysed the mitotic index in apices of *Datura stramonium* at various stages of development. In this species, as already mentioned, the apex enters the transitional stage as soon as 7 or 8 leaf primordia have been formed. The apex was divided into 4 zones, partially based on staining characteristics. These zones, which could be recognized in both the vegetative and transition apices (Fig. 6.13), were

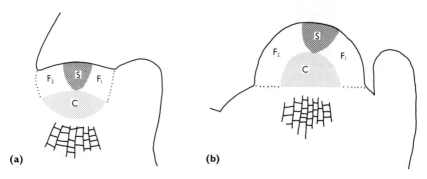

Fig. 6.13 Apical zonation in *Datura stramonium*. C, central zone; F_1 and F_2, flank zones; S, summit zone. The rib meristem is also indicated. (a) Vegetative apex, from a plant which had formed 5 leaves and leaf primordia. (b) Transition apex in stage 4, from a plant with 7 or 8 leaves and leaf primordia. (Redrawn after Corson,[111] Fig. 2, p. 1128.)

designated the summit zone (S), central zone (C) and two peripheral flank zones (F_1 and F_2), as seen in longitudinal section. A pith rib meristem was present below the central zone. When the mitotic indices of the various zones were analysed by counting the percentage of nuclei in stages of mitosis, it was found that in the vegetative apex the mitotic index of the

Table 6.1 Average mitotic indices for the apical zones of the vegetative and transition apex of *Datura stramonium*. C, central zone; F_1 and F_2, flank zones; S, summit zone. (From Corson.[111])

State of apex	F_1	S	F_2	C
Vegetative	3·69	1·98	3·87	4·34
Early-transition	3·74	2·88	4·33	6·02
Late-transition	3·86	3·27	4·67	5·21

summit zone was significantly lower than that in the other zones, which did not differ significantly from one another in this respect (Table 6.1 and Fig. 6.14). In the early and late transitional apices the mitotic index of both the

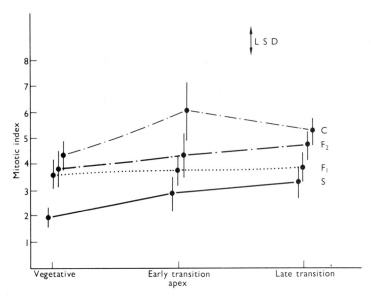

Fig. 6.14 Changes in the mitotic index in the various apical zones of *Datura stramonium* from vegetative to late-transition stages. Vertical lines indicate twice the standard error on each side of the mean. LSD, least significant difference between different zones. C, central zone; F_1 and F_2, flank zones; S, summit zone. (From Corson,[111] Fig. 7, p. 1130.)

summit and the central zones increased significantly (Table 6.1 and Fig. 6.14). By using the drug colchicine to arrest mitosis at the metaphase stage, it was possible to calculate the rate of cell division in the various zones, and to demonstrate that in this species the mitotic index is a measure of the rate of cell division. In *Datura* the increases in mitotic index in the various zones paralleled increases in RNA, phospholipid and total protein at transition to flowering, as shown by histochemical methods. As Corson[111] points out, the mitotic index is highest at a stage of development corresponding to elongation and doming of the apex. In *Sinapis*, also, an increase in DNA synthesis and mitotic activity was shown in the central axial zone at the pre-floral stage, and a subsequent decrease as the apex entered the floral stage.[39] The same is true in *Perilla*.[587] This increase in mitosis in the central region thus seems to be correlated with a change in form in the apex, as might be expected; whether it is obligatorily correlated with the flowering process, however, is rather less clear, especially since partial reversion to the vegetative phase can occur in *Sinapis*,[39] and the apex of *Perilla* never forms a terminal flower.[587] Corson[111] points out that although there is less mitotic activity in the summit zone of *Datura* during the vegetative phase of development, some divisions do occur; and since the number of cells in this region remains constant their daughter cells must contribute to the other zones. Thus this region, although less mitotically active, is not a true *méristème d'attente*, since it participates in histogenesis in the vegetative phase.

The view that the central axial zone is a cytologically and physiologically distinctive region of the apex, however, is supported by experiments in which it is claimed that treatment with the analogue thiouracil, which inhibits DNA synthesis, inhibited flowering by specifically suppressing any increase in DNA synthesis in the central zone.[311] Also, in *Rudbeckia* a stimulation of DNA synthesis was obtained only in the central zone, after treatment with gibberellin.[39] In *Perilla* also the floral stimulus specifically affected mitotic activity, RNA content and number of ribosomes in the cells of the central zone.[587] In *Sinapis*, one of the earliest results of floral induction is the release of nuclei from the G_2 phase of mitosis.[40] These nuclei have thus synthesized DNA but are apparently blocked from mitosis. The G_2 phase is a period of active RNA and protein synthesis.

Since the *anneau initial/méristème d'attente* theory was first put forward, it has stimulated much work on cell division and DNA synthesis in the shoot apex. It is becoming increasingly evident that there are great differences between species in the distribution of mitotic activity in the apex. For example, using ^{3}H-orotic acid, a precursor of DNA, followed by histo-autoradiography, Knox and Evans[314] found that the density of silver grains in the exposed film, indicating the site of incorporation of the labelled substance, was highest at the summit of presumably vegetative apices of the LDP *Lolium temulentum* maintained in SD (Fig. 6.15). There was thus no evidence of a *méristème d'attente*. In LD, activity increased in all parts of the apex (Fig. 6.15). By contrast, counts of the distribution of silver grains in autoradiographs of vegetative apices of *Helianthus annuus*, which had been grown in sterile liquid medium and supplied with ^{3}H-thymidine, showed very little difference between the number of grains overlying nuclei in the central zone and the value for background radiation. In the peripheral region the counts were much higher. No mitotic figures were observed in the central zone. At the outset of flowering, labelling was uniform throughout the apex.[493] These workers conclude that DNA synthesis must occur at a very slow rate in the cells of the central zone, if indeed it occurs at all; or that activity in this region must occur only periodically. Thus in this species there seems to be a true *méristème d'attente*. The

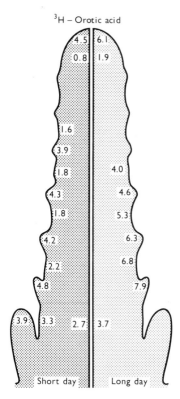

^3H – Orotic acid

Short day   Long day

Fig. 6.15   Pattern of incorporation of ^3H-orotic acid, as grains per 10·9 μm^2, in median L.S. of apices of *Lolium temulentum*. The left-hand side of the apex gives average values for vegetative apices, the right-hand side values for induced Day 2 apices. (From Evans, L. T., *in* Evans,[171] pp. 328–349, Fig. 14–9a, p. 346. After Knox and Evans,[314] Fig. 2a, p. 1088.)

reason for the striking discrepancy between the observations on apices of *Lolium* and of *Helianthus* may perhaps be related to the shapes of these apices in the vegetative phase. *Lolium*, like other grasses, has an elongated, conical apex, whereas that of *Helianthus* is almost flat. At the transition to the floral phase, there is little change in shape of the grass apex, but a great increase in size and some change in shape in the sunflower. The involvement of the cells of the central zone at various stages of development may thus be related to geometrical considerations. It is, however, very important to reach some understanding of the factors controlling cell division and its distribution in the shoot apical meristem, and further experiments and observations must be directed to that end.

FORMATION AND DEVELOPMENT OF LATERAL ORGANS

Organogenesis

During and after the transition to the flowering phase the inflorescence apex may give rise to a series of small, leaf-like bracts. Subsequently florets are formed. In the apex of an individual flower, the various lateral members are formed in sequence. Organogenesis is well demonstrated in a Ranalian

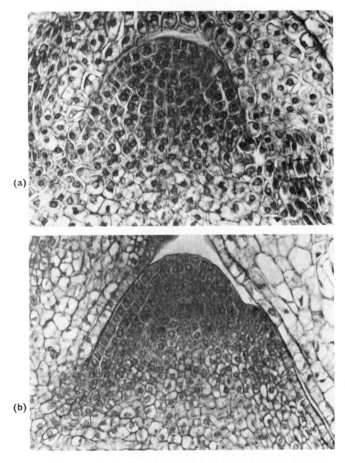

Fig. 6.16 L.S. apices of *Aquilegia formosa* var. *truncata*. (**a**) Vegetative apex. ×392. (**b**) Transitional apex, showing changes in height and width.   ×352. (From Tepfer,[500] Plates 48a and 50b, pp. 581 and 585. Originally published by the University of California Press; reprinted by permission of the Regents of the University of California).

flower such as *Nuphar* or *Aquilegia*, since many floral organs of each category are formed. In *Aquilegia* the various lateral organs of the vegetative and reproductive phases—in this case leaf, bract, sepal, petal, stamen, staminodium and carpel—have a comparable origin in the second and third layers of the apical meristem, but soon display differences in form and manner of growth.[500] Stages in ontogeny of the flower of *Aquilegia* are shown in Figs. 6.16 and 6.17. Sepals, petals, stamens and carpels also

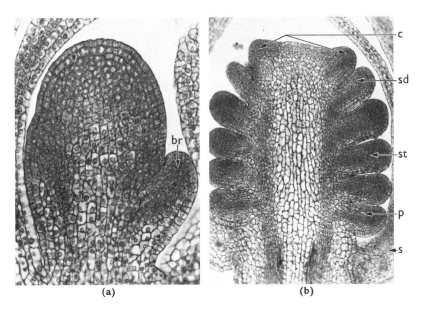

(a)   (b)

Fig. 6.17 L.S. developing floral apices of *Aquilegia*. (a) Transitional apex forming bracteoles (br).   × 276. (b) Developing flower with sepal (s), petal (p), stamen (st), staminodium (sd) and carpel (c) primordia.   × 152. (From Tepfer,[500] Plates 51b and 56b, pp. 587 and 597. Originally published by the University of California Press ; reprinted by permission of the Regents of the University of California).

originate by periclinal divisions in the second layer of the apex in the flower of *Portulaca*.[485] Observations of this kind support the view that these organs are all lateral members of the shoot apex, and can be interpreted as different manifestations of a leaf-like structure.

In the epigynous flower of *Vinca rosea*,[48] the **sepals** are formed in spiral sequence. The apex then enlarges rapidly and forms 5 **petal** primordia simultaneously. In this gamopetalous flower the bases of the petals unite during ontogeny and form the upper part of the corolla tube; the lower part of the tube is formed by zonal growth in the common basal regions of both petals and **stamens**. The union of petals and stamens can be seen in transverse sections of the developing flower (Fig. 6.18).

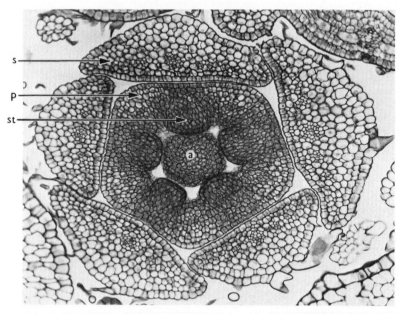

Fig. 6.18 T.S. young flower of *Vinca rosea*, showing the floral apex (a) and sepal (s), petal (p) and stamen (st) primordia.   × 205. (From Boke,[48] Fig. 8, p. 416.)

The diameter of the floral apex is considerably reduced by the end of stamen formation. Primordia of sepals, petals, and **carpels** are formed by periclinal divisions in the second tunica and outer corpus layers.[49] In early stages of development, stamen primordia are not unlike those of petals (Fig. 6.19a). Later, intercalary growth at the base of the stamen primordium gives rise to a short region without sporogenous cells, the *filament*, as distinct from the terminal **anther** portion. The same occurs in the stamens of *Downingia*.[302] During the development of the anther, the hypodermal layer divides to give rise to the sporangium wall (other than the epidermis), part of the **tapetum**, and the **sporogenous tissue** (Fig. 6.20a and b). The sporogenous tissue will later divide to form the pollen grains or microspores, and the tapetum is a layer of cells that supplies metabolites to the sporogenous cells.

After the formation of the stamens, the diameter of the floral meristem increases again, and the cells at the periphery become raised up to form a meristematic ring. Carpel primordia are formed at two opposite sites in

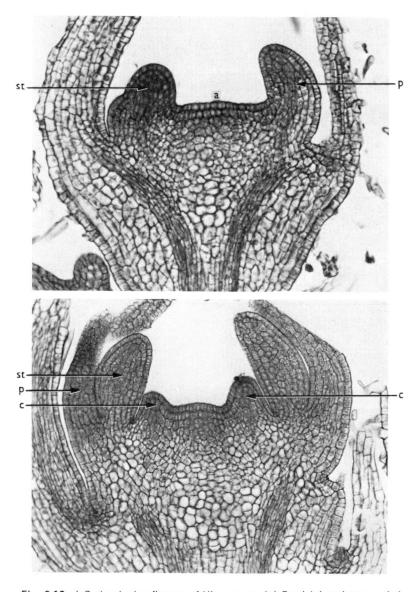

Fig. 6.19 L.S. developing flowers of *Vinca rosea*. (a) Petal (p) and stamen (st) primordia have been formed by the floral apex (a).   × 230. (b) A slightly later stage; carpel (c) primordia have now been formed.   × 196. (From Boke,[49] Figs. 3 and 15, pp. 536 and 540.)

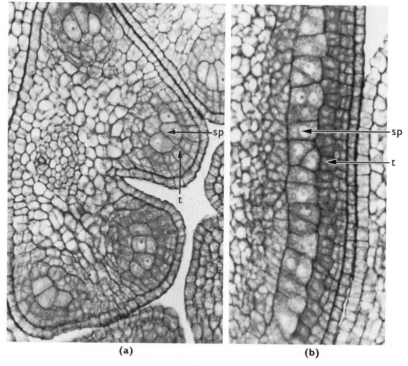

(a)   (b)

Fig. 6.20 Developing stamens of *Vinca rosea*. (a) T.S. stamen, showing the developing sporogenous cells (sp) and tapetum (t).   ×395. (b) L.S. part of an anther, showing the developing sporogenous tissue (sp) and tapetum (t).   ×410. (From Boke,[49] Figs. 10 and 13, pp. 538 and 540.)

this ring (Fig. 6.19b). Later, carpellodes (sterile members) are formed from other parts of the meristematic ring.[49]

A careful investigation of the growth of the lateral organs in the flower of *Downingia* shows that, like leaves, the floral members exhibit apical, marginal and sometimes adaxial growth. Sepals undergo apical, marginal and intercalary growth, as do the petals, but in these apical growth is of shorter duration, while there is more extensive activity of the plate meristem.[301] Even in the stamen primordia, which are thicker and have much broader apices than the primordia of sepals and petals, some marginal growth can be detected. The carpels also exhibit marginal growth, and possess an adaxial meristem which contributes to growth in thickness.[302] Thus, not only in their origin, but also in their mode of growth, the lateral organs of the flower can be compared with leaves, though as explained earlier they are doubtless records of a different physiological stage of the parent apex. Some writers believe that stamens, in particular, show more parallels with stem axes than with leaves, partly because they have a more deep-seated origin,[446] and there is no doubt that lateral members of the flower can be more easily compared with leaves in some species than in others. It is perhaps more stimulating to consider them developmentally as expressions of different physiological states of the apex.

Development of the stamen and pollen

Recently there has been a considerable amount of work on the development of the stamen and the pollen grains, using not only classical anatomical techniques but also histochemistry, autoradiography and both transmission and scanning electron microscopy. The scanning electron microscope is an instrument which, like the more familiar transmission electron microscope, bombards the specimen with electrons, but it gives greater depth of focus. Whole specimens rather than sections are observed, and observation is restricted to surface features.

The developing *pollen mother cells* and the *pollen* to which they give rise are of particular cytological interest for a number of reasons: for example, the pollen mother cells are a site of meiosis in higher plants, and their nuclear divisions are often synchronized; also, the pollen grain is the bearer of the male gametes, and the form of the pollen wall is of diagnostic significance.

The young anther usually becomes 4-lobed at an early stage of development and the hypodermal cells become distinctive in each lobe (Fig. 6.19). This tissue, the *archesporium* or archesporial tissue, then divides periclinally, forming a primary parietal layer towards the outside and a primary sporogenous layer towards the inside.[537] The primary parietal layer divides again to form a variable number of cell layers composing, from the outside in, an *endothecium*, below the epidermis, one or more middle layers, and the *tapetum*, which is usually a single layer of cells (Fig. 6.21). According to the species, the *primary sporogenous cells* may either function directly as pollen mother cells (pmc), or may divide mitotically to form a larger number of sporogenous cells. The tapetum is closely involved in development of the pollen grains, and is the source of important metabolites. The cells of the tapetum may be multinucleate or endopolyploid, in different species. Each pollen mother cell divides by meiosis to give rise to a tetrad of *microspores* or *pollen grains*. Changes in cellular metabolism during these developments have been studied by various techniques. In some instances it is not yet known whether some of these observations, made on individual species, are of general significance. However, since many of them are of great interest they are discussed here.

Histochemical studies have shown that in anthers of *Zea mays* the concentration of RNA and of protein falls from the time the pmc's are

vb

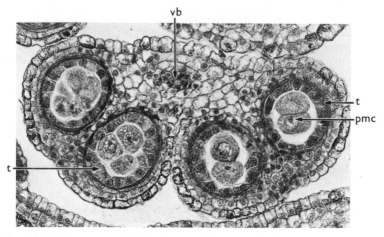

Fig. 6.21 T.S. stamen of *Cabomba caroliniana*, showing the two-lobed anther with 4 pollen sacs. The vascular bundle (vb) in the connective region can be seen. pmc, pollen mother cells; t, tapetum.   × 250.

forming until during meiosis. There was no increase during meiosis.[371] In the anthers of the paeony, *Paeonia*, however, the cytoplasmic RNA and protein increased in the archesporial cells and pmc's prior to meiosis. During meiosis the content of RNA and protein in the cytoplasm of the tapetal cells increased, reaching a maximum at the time of separation of the tetrads.[449] The pmc's lose most of their cytoplasmic RNA and protein at the time of microspore formation, but after mitosis in the microspore these substances increase again; at the same time the tapetum breaks down.[449] Similar results were obtained by counting the number of silver grains in autoradiographs of anthers supplied with ^{3}H-cytidine or ^{3}H-leucine, respectively precursors of RNA and of proteins. It is suggested that proteins or their precursors are probably supplied to the pmc by the tapetum.[448]

A good deal of interest has centred on the interrelationships of the developing microspores with each other and with the tapetum. In *Lilium*, at least, during prophase of meiosis the pmc's are connected by massive strands of cytoplasm, sometimes called **cytomictic channels**. These are severed at some time during meiosis, and an imperforate wall of the carbohydrate callose, the 'special wall', is formed around the tetrad of spores. It is suggested that the existence of the cytomictic channels effectively means that the pmc's share a common cytoplasm, resulting in the synchronous behaviour of their nuclei. Moreover, the differences that the pollen grains may subsequently express, such as the manifestation of different incompatibility systems, may be attributable to the impervious nature of the **callose wall**, so that subsequent development takes place in relative isolation.[255] On the basis of work with the light microscope with pollen of *Arnebia*, it is claimed that nuclear material migrates through the cytoplasmic channels.[21] In species of orchids which form pollen grains in aggregates or massulae, cytoplasmic connections persist until the pollen matures, and in these species the mitotic division in the pollen grains is also synchronous.[252] The results of experiments in which ^{14}C-thymidine was supplied to developing flowers of *Lilium* seem to support the view that the callose wall is rather impermeable. The labelled material moved freely into the sporogenous cells up to the pachytene stage of meiosis, but after this stage it failed to penetrate into the pmc's until the tetrad wall broke down. The spores were able to take up the tracer after their release. It is suggested, therefore, that the callose wall may function in regulating the movement of materials, particularly macromolecules, in the pollen sac.[255] The callose wall also seems to exert some restraint on the spores, since they increase considerably in volume after release from the tetrad.[371] The callose is thought to be synthesized by dictyosomes in the pmc's.[157]

During early stages of formation of the pollen grains, small bodies, the pro-Ubisch or pro-orbicular bodies, are formed in the tapetal cells, sometimes towards the inner face of these cells (Fig. 6.22a). When the callose wall of the tetrads breaks down, so also do the walls of the tapetal cells, and the pro-orbicular bodies are extruded into the pollen sac (Fig. 6.22c). At this stage they become irregularly coated with **sporopollenin**, a substance which forms the outer wall of the pollen grain, and are now termed **Ubisch bodies** or **orbicules**. Deposition of sporopollenin continues on both the orbicules and on the outer wall of the developing pollen grains; it is suggested that this substance may be synthesized from materials released into the pollen sac by the breakdown of both the special callose wall and the tapetal cell walls.[156, 253] Thus precursors may be synthesized both in the young microspores and in the tapetal cells. In *Lilium*, a second set of spherical bodies is formed in the tapetal cells shortly after the pro-orbicular bodies. These are concerned with the synthesis of *Pollenkitt*, an adhesive pigmented material which coats the mature pollen grain.[250]

At about the time that the tapetal cells break down and extrude the orbicules into the pollen sac, the cells of the **endothecium** expand in a radial direction and radial bars of thickening form on the inner side of the cell wall. In different species, these may form a variety of patterns.[434] In *Chenopodium*, these thickenings were found to consist mainly of α-cellulose, and lignin was not detected in various histochemical tests.[183] In some other species, at least, the thickenings are usually considered to be lignified, but perhaps further study on this point is needed. Fossard[183] suggests that during much of anther development a product of the tapetum inhibits the development of the endothecium, and that it is only after the

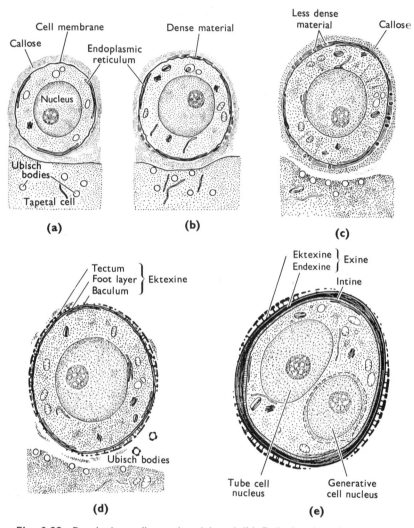

Fig. 6.22 Developing pollen grains. (a) and (b) Endoplasmic reticulum (ER) forms in the cytoplasm, delimiting the future position of the pores. Dense fibrous material coats the cell in regions where the ER is absent. Orbicular or Ubisch bodies form in the adjacent tapetal cells. In (a)–(c) callose surrounds the pollen grain cell. (d) Callose decreases and part of the exine, the ektexine, develops. The orbicular or Ubisch bodies are freed from the tapetum. (e) The endexine and intine are formed around the mature pollen grain, and the nucleus divides to form a vegetative or tube cell nucleus and a generative cell nucleus. (From Echlin,[155] Figs. 5–9, pp. 86 and 87. Copyright (1968) by Scientific American, Inc. All rights reserved.)

cessation of sporopollenin synthesis and breakdown of the tapetum that the endothecium can develop. This tissue is functionally important, since it participates in the eventual dehiscence of the anther (Fig. 6.23). The endothecium does not form all round the periphery of the pollen sac, but a

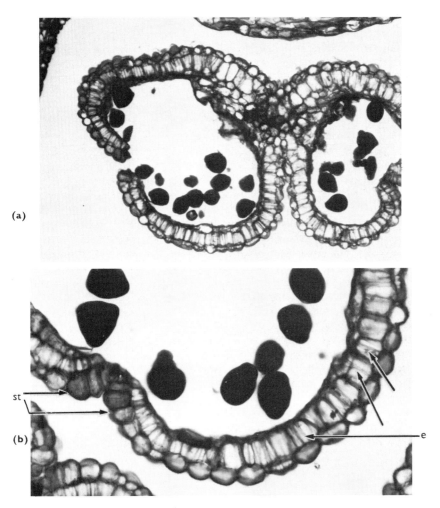

Fig. 6.23 T.S. anther of *Prunus* sp. (a) Showing dehiscence. Pollen grains are present in the pollen sacs.   × 150. (b) Enlarged view of a region of the anther wall, showing the endothecium (e) with its bars of thickening (arrows) and the thin-walled cells of the stomium (st).   × 300. (Slide by courtesy of Dr. G. L. Webster.)

strip of thin-walled cells, the **stomium**, remains. The latter usually either runs vertically along the length of the anther, between the two locules of each anther lobe, or forms a peripheral strip near the top of the anther lobe. Dehiscence occurs in the region of the stomium; the precise mechanism would be worthy of further study. Moreover, the factors underlying the differentiation of a different kind of tissue, the stomium, in certain well defined regions in the anther, require investigation.

The development of the highly specialized wall of the pollen grain has been the subject of much work. Pollen is, of course, biologically very important, being the bearer of the male gametes. It also constitutes a health hazard to some individuals, who are allergic to the pollen of particular species and suffer from hay fever. Being light, many pollen grains are present in the atmosphere; they can be carried distances of at least 400 miles, probably much further. Because of the nature of their wall, pollen grains are extremely resistant to decay, and fossilized pollen grains have been found in deposits some 100 million years old.[155] Often fresh and fossilized pollen appear very similar, even when viewed with the scanning microscope.[351] Because of this excellence of preservation, coupled with the specific structural features of the pollen wall, which often enables the identification of the pollen grain down to the species level, pollen is used extensively by ecologists. The nature of the former vegetation of a particular location can often be established from the pollen found in the peat deposits. This is rendered possible partly because of the vast quantities of pollen that some plants produce. For example, the spruce forests of southern and central Sweden are said to produce 75,000 tons of pollen per year;[155] considering the weight of a pollen grain, this is a prodigious amount.

While the pollen grains are still within the special callose wall, the spore wall called the **primexine** is formed. This wall is cellulosic.[158] In certain regions of the periphery of the young spores strips of endoplasmic reticulum are laid down just beneath the plasmalemma (Fig. 6.22a, b); no wall is formed in these areas, which constitute the sites of the future **pores**.[157, 248] In some species, a pore or furrow forms at the point of contact in the tetrad, or towards the outside of the tetrad. The pores represent thinner regions in the wall through which the pollen tube may emerge on germination. The number of pores or furrows is very constant for each species. The genetic control over this pattern in the primexine must be attributable to the haploid nucleus of the spore.[248]

The **exine** is then formed. Precursors of the column-like bacula begin to project from the surface (Fig. 6.22d). The upper and lower ends of the bacula then spread laterally to form a perforated roof, sometimes called muri, and a floor. The floor, pillars and roof constitute the ektexine (Fig. 6.22e), which then becomes impregnated with sporopollenin. During this time the callose wall is breaking down, and the pollen grains are set free.[155, 251] Thus the distinctive pattern of the exine develops while the microspores are still within the callose wall and, if this is impermeable, it follows that the pattern must be determined by the haploid nucleus.[250] The endexine, part of the outer coat or exine, forms below the ektexine, and finally the cellulosic inner wall, the *intine*, forms within this (Fig. 6.22e). Godwin[214] has pointed out that if a morphogenetic system controlling these events is recognized, it may be susceptible to experiment. A genetically controlled system which determines such complex patterns would clearly be of great interest.

At about this time the small vacuoles in the cytoplasm of the pollen grain coalesce to form a larger one, and the nucleus divides to give the *generative* and *vegetative nuclei* (Fig. 6.22 and 6.24). In the pollen grain

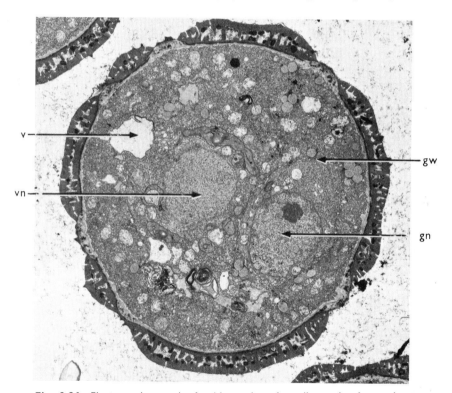

Fig. 6.24 Electron micrograph of a thin section of a pollen grain of sugar beet, *Beta vulgaris*. The pollen grain is binucleate at this stage, with a vegetative nucleus (vn) and a generative cell with nucleus (gn) and cytoplasm surrounded by a wall (gw). v, vacuole.   × 5000. (From Hoefert,[260] Fig. 11, p. 367.)

cytoplasm a structure known as a reticulum complex, composed of endo-plasmic reticulum and associated ribosomes, is present.[260] The division which occurs in the pollen grain is an asymmetric one, and the smaller generative cell is separated from the vegetative nucleus and cytoplasm by a curved wall (Fig. 6.25).[14] This wall resembles that resulting from similar unequal divisions in other parts of the plant (compare Figs. 2.9 and 5.28). In orchids, the two nuclei resulting from the division differ in shape. There are no plasmodesmata between the vegetative and generative cells.[252] The exine of the pollen grain is now sculptured in various distinctive ways (Fig. 6.26). One or more furrows or colpi may be present, and germ pores can be discerned under the light and scanning electron microscopes (Fig. 6.26).

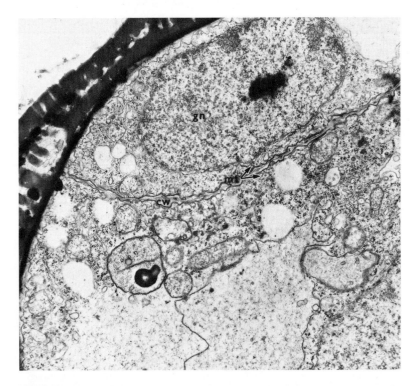

Fig. 6.25 Electron micrograph of part of a pollen grain of sugar beet, showing the curved wall (cw) delimiting the generative cell with its nucleus (gn) and cytoplasm from the rest of the pollen grain. The dark region on the generative nucleus is a particle of stain. mt, microtubules, adjacent to the cell wall.   ×13,780. (By courtesy of Dr. L. L. Hoefert.)

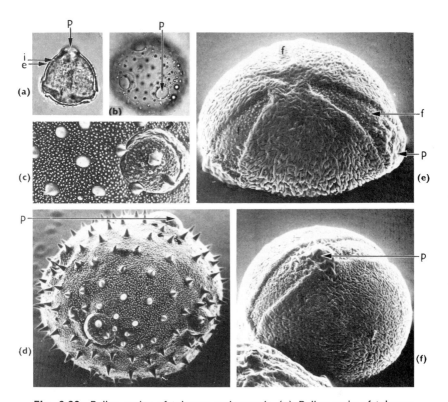

Fig. 6.26 Pollen grains of tobacco and squash. (a) Pollen grain of tobacco, *Nicotiana tabacum*, seen in optical section under the light microscope, showing the exine (e) and intine (i) and the three germ pores (p). Compare (e) and (f). ×485. (b) Pollen grain of squash, *Cucurbita pepo*, as seen under the light microscope. Several pores (p) and some of the ornamentation on the exine can be discerned. Compare (c) and (d).   ×175. (c) Part of the exine of a pollen grain of squash, as seen under the scanning electron microscope. Part of the exine in the region of the pores apparently lifts off; pointed projections of two sizes are present on the exine.   ×775. (d) Whole pollen grain of squash as seen under the scanning electron microscope. Note the pores (p) and spine-like projections on the exine. Compare (b).   ×445. (e) and (f) Pollen grains of tobacco as seen from different angles under the scanning electron microscope. The approximately tetrahedral grain has 3 germ pores (p) each of which occurs in a colpus or furrow (f). The sculptured exine is split in the region of the pores. Compare (a).   ×1710. ((c)–(f) from photographs by M. J. York.)

Later, the generative nucleus divides to give rise to two ***sperm nuclei***. On reaching an appropriate stigmatic surface, the pollen grain germinates, putting out a ***pollen tube*** through one of the germ pores (Fig. 6.27).

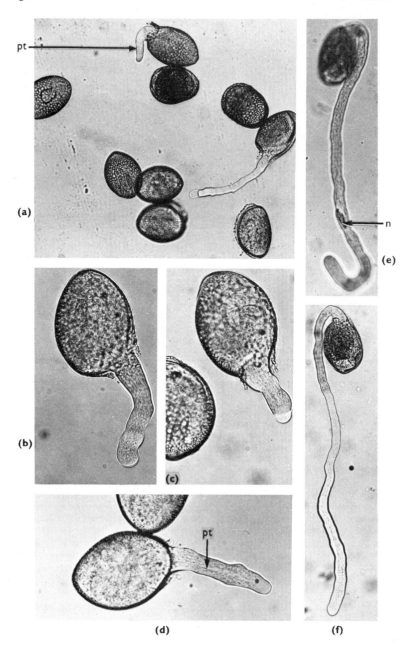

Recent work in which the synthesis of various substances by the pollen grain and stigma has been demonstrated suggests that the relationship between these organs is a complex one. Extracts of the stigmas of 10 different species revealed the presence of several compounds. It is considered that the phenolic compounds may interact with growth substances to control the germination and growth of the pollen, and may perhaps account for the specificity of stigmas which allow only certain pollen grains to germinate.[352] Discovery of a number of hydrolytic enzymes in the intine of the pollen of 10 species suggests that they, in turn, may perhaps aid penetration of the stigma.[315] On reaching the ovule, fertilization is achieved, one sperm nucleus fusing with the egg nucleus and the other with the fused polar nuclei. These events are discussed further in Chapter 8.

While these important events are taking place in the anther, the filament

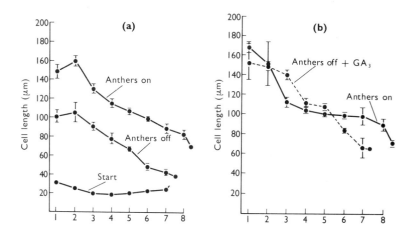

Fig. 6.28 Experiments on elongation of stamen filaments of *Nigella hispanica*. (a) Effect of removing anthers on epidermal cell elongation in the filaments. On the horizontal axis groups of 10 cells along the length of the filament are indicated, 1 being the basal group. 'Start' shows the cell length of filaments at the time of emasculation. 'Anthers on' shows the mean cell length in intact filaments at anthesis. 'Anthers off' shows mean cell length in emasculated filaments from the same flower. (b) Effect on cell length of gibberellic acid (GA_3) in lanolin applied to tips of anther-less filaments. (From Greyson and Tepfer,[220] Figs. 3 and 8, pp. 973 and 975.)

Fig. 6.27 Germinating pollen grains of *Amaryllis* sp., growing in a 10 per cent sucrose solution. (a)–(d) After 1 hour in the solution. Pollen tubes (pt) have emerged through a pore. (a) ×150; (b)–(d) ×300. (e) and (f) After 5 hours in sucrose. The pollen tube is quite long. n, nucleus. ×150.

of the stamen is elongating. Some such mechanism would be necessary to ensure access of the pollen to the stigma in many self-fertilized species. In *Nigella* the filament undergoes a 10-fold increase in length in 16 days, by means of both cell division and cell enlargement. A diffuse intercalary meristem is present, its main activity being in a nearly central position.[219] In experiments in which the anthers were removed, the filaments attained only 59 per cent of the length of filaments in intact stamens, but this could be increased to 87 per cent by application of GA (Fig. 6.28).[220] The effects of this hormone seemed to be on cell enlargement rather than cell division, so that another factor may also be involved.

Development of the carpels and ovules

As we have seen, in most species the **carpels** originate as lateral organs on the floral meristem, often in a manner similar to that of leaves, bracts and other floral organs.[500, 524] In most species the floral meristem is not entirely used up in the formation of the carpels,[49] but in a few species a single carpel is formed occupying an apparently terminal position. The case of *Drimys lanceolata*, with a single terminal carpel, is interesting in this respect, since the related species *D. winteri* has several lateral carpels.[524, 528, 529] The marked difference in form at maturity of the carpels of these species is attributed to early differences in cellular activity. In the carpels of most species marginal and often adaxial growth occurs.[49, 154, 302]

Carpels often have three vascular traces, one dorsal and two ventral strands, though neither this nor their arrangement is by any means invariable.[525, 527] Usually the ventral strands supply the ovules.

In some species, the carpels remain separate structures, and the ovule-bearing part of the plant, the **gynoecium**, is said to be **apocarpous**. In other species, union of the carpels occurs in various ways, and the gynoecium is **syncarpous**. The carpel can be considered to comprise three parts, the basal, ovule-containing part, called the **ovary**, and the **style** and **stigma**. The stigma is the region on which the pollen is deposited; the pollen tube grows down the style to the ovary.

Even in species with a single carpel, such as a pea pod, the margins of the carpel join together to form a suture. In syncarpous gynoecia, fusion may occur only in the ovary region, leaving free styles and stigmas, or these may also unite into a more or less single structure. Syncarpy may be the result of zonal growth, when the independent growth centres have merged at an early stage of development and the whole perimeter has engaged in vertical growth (congenital fusion), or fusion between the margins of separate carpels may occur in various ways during ontogeny (post-genital fusion).[121] If the carpels folded before becoming joined, the ovary has several cavities or **locules**; if not, the ovary is unilocular.

The ovules are borne on a **placenta**, a region of tissue situated where the margins of a single carpel, or more than one, have fused. The two ventral bundles from the carpellary margins fuse and supply the placenta and ovules. According to the position of the placentae, various types of placentation may occur (Fig. 6.29). The number of placentae is usually

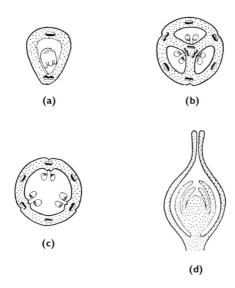

(a)   **(b)**

(c)

(d)

Fig. 6.29 Types of placentation. (a) Marginal. (b) Axile. (c) Parietal. (d) Basal. (From Puri,[417] Table I (in part), p. 605.)

the same as the number of carpels. In apocarpous gynoecia, the placentation is sometimes described as **marginal**. In syncarpous gynoecia, placentation is **axile** if the placenta occurs on the fused margins of the same carpel, and derives its vascular supply from traces to that carpel (Fig. 6.29b); it is **parietal** if the placenta is borne on the fused margins of different carpels and derives its vascular supply from different carpels (Fig. 6.29c).[417] These distinctions are not always rigid, however. In some families, ovules are borne on a column of tissue arising from the base of the ovary (**free-central** placentation), or a single ovule occurs on the base of the ovary (**basal** placentation; Fig. 6.29d). These types are believed to have been derived from axile placentation.[154, 417]

The stigma usually has rather large cells which may have a secretory function; the epidermis often bears papillae. In the centre of the style specialized tissue occurs, specialization of a region of papillose epidermis.[154] This tissue is variously called conducting or **transmitting tissue**, and it constitutes the path of the pollen tube. It may extend up the stalk or funicle of the ovule. These specialized regions of the style and stigma apparently hold the clue to some types of incompatibility. For example, in some species or varieties the stigma may have a covering of cutin; the pollen may lack the necessary enzymes to break down the cutin, and be unable to reach the style.[353] A comparative study of pollen growth in self-compatible and self-incompatible populations of a crucifer led to the conclusion that the stigma possessed a growth substance, which, if present in sufficient amount, could promote the adherence of the pollen grains to the stigma and also stimulate germination and growth of the pollen tube.[331] An ultra-structural study of the transmitting tissue in the style of the cotton flower[283] has shown that in one region of the style the pollen tube actually makes its way through a region of the cell walls, not between cells. These cell walls are somewhat porous, and the pollen tube evidently follows the path of least resistance down the style. In *Lilium*, the central canal cells have wall ingrowths and resemble transfer cells.[432a]

Germination of pollen can be promoted by sucrose solutions, at least in some species (Fig. 6.27), and this has facilitated experiments in which ovules were artificially pollinated. Tobacco ovules with part of the placenta were cultured under sterile conditions, and either transferred to a culture of growing pollen tubes or left *in situ* and supplied with growing pollen tubes. Viable seeds were obtained.[23] Earlier, ovules with portions of placenta of the opium poppy, *Papaver somniferum*, had been cultured and pollen dusted over the ovules and on the medium. The pollen germinated, and double fertilization ensued.[299] These techniques may be valuable to plant breeders, in cases of sterility where the reason for incompatibility is to be found in the stigma, style or ovary, since these structures are effectively by-passed.

The development of the ovule and the process of fertilization will be described briefly by taking a particular example, that of the Composite guayule, *Parthenium argentatum*.[164] In this species a single ovule is present basally. The ovule is initially upright, but since growth is greater on the upper side the ovule gradually becomes inverted. This type of ovule is called **anatropous**. Various types of ovule occur in other species. Ovules which remain vertical are called **orthotropous**, while those which curve partially, so that the embryo sac itself is curved, are called **campylotropous.**

The ovule of guayule has a **megaspore mother cell** in the hypodermal layer, at one end of the **nucellus** (Fig. 6.30a, b). There is only one **integument** in this species; in many plants two integuments develop. The stalk of the ovule, by which it is attached to the placenta, is called a **funicle** or funiculus. The megaspore mother cell undergoes meiosis and gives rise to 4 **megaspores** (Fig. 6.31a). The megaspore at the chalazal (basal) end is the largest and becomes the **embryo sac**; the 3 megaspores at the micropylar end become crushed (Fig. 6.31b). Kaplan[303] has recently shown that

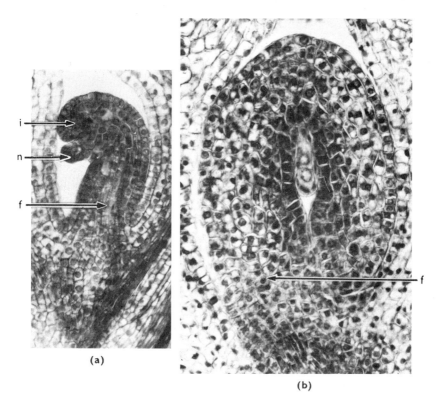

Fig. 6.30   Developing ovules of guayule, *Parthenium argentatum.* (a) The ovule, attached by the funicle (f) or stalk, is still incompletely inverted. The megaspore mother cell is evident in the nucellus (n), with the single integument (i) above. × 290. (b) The anatropous ovule is now inverted. The megaspore mother cell has divided to form a dyad.   × 340. (From Esau,[164] Plates 1A and 5A, pp. 105 and 109.)

in *Downingia* the first meiotic division of the megaspore mother cell is an unequal one, giving rise to a larger cell at the chalazal end. In the second division the chalazal cell maintains its distinction by undergoing division slightly before the micropylar derivative. In view of the importance of unequal divisions in processes of differentiation in general, this establishment of polarity during the division of the megaspore mother cell may be important in determining the survival of the chalazal megaspore.

Following vacuolation of the embryo sac in guayule the nucleus divides (Fig. 6.31c), and the two daughter nuclei move to opposite ends of the cell. As a result of enlargement of the embryo sac the nucellar epidermis becomes crushed, and the embryo sac lies in a cavity protected by the integument.

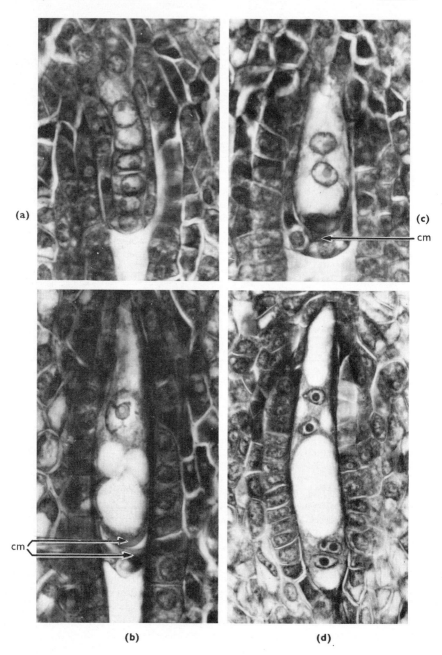

(a)

(b)

(c)

cm

cm

(d)

Nuclear division occurs in the embryo sac to give rise to 4 (Fig. 6.31d) and then to 8 nuclei. Three of these are *antipodal* nuclei, at the chalazal end of the embryo sac; two are *synergid* nuclei, at the micropylar end; two are *polar nuclei* and one is the *egg nucleus*. After the division into 8 nuclei, cell walls are formed and 7 cells are present in the fully differentiated embryo sac: the 3 antipodals, the 2 synergids, the egg and the large central cell, situated centrally and containing the fusion or secondary nucleus resulting from the fusion of the 2 polar nuclei.[164]

In the great majority of plants, at fertilization the two sperm nuclei of the pollen tube enter through the micropyle and penetrate the embryo sac. One fuses with the egg nucleus, and the resulting diploid *zygote* later develops into the *embryo*; the other fuses with the fusion nucleus, to form a triploid nucleus, the *primary endosperm nucleus*, which divides to form the *endosperm*. The fate of the cytoplasm of the sperm cells has not yet been established. These events are considered further in Chapter 8.

The fertilized ovule develops into the *seed*, the integument(s) becoming the testa or seed coat, the nucellus (in the relatively few species where it persists) becoming a nutritive tissue called perisperm, and the endosperm and embryo develop as noted above. The ovary becomes the *fruit*. The structure of mature fruits and seeds is considered in the following chapter, and the development of the embryo in Chapter 8.

Sex expression

In some species, various treatments can affect the development of stamens or carpels, and thus the 'sex' of the flower. These experiments are interesting, since they apparently invoke switching mechanisms in development. This may occur at the level of the floret primordia themselves, or of their lateral organs. In species of *Carex*, for example, a fairly regular pattern of formation of male or female flowers occurs, differing in each species. In the inflorescence of *Carex*, two kinds of meristems occur initially: male flower primordia and female spikelet primordia. The latter can develop either as a lateral spike or as a female flower. Treatment with IAA, TIBA and kinetin indicated that in the presence of high auxin and high kinetin levels lateral spikes tend to develop; with high auxin and low kinetin, female spikelets occur; and with low auxin and low kinetin male flowers

---

Fig. 6.31 L.S. parts of the nucellus and integument in guayule (see also Fig. 6.30). (**a**) A linear tetrad of megaspores (4) has been formed by division of the megaspore mother cell. (**b**) Uninucleate embryo sac; the remains of the 3 crushed megaspores (cm) can be discerned. (**c**) Binucleate embryo sac with crushed megaspores. (**d**) Tetranucleate embryo sac. A single layer of nutritive cells borders the embryo sac. (**a**)–(**c**) ×890; (**d**) ×560. (From Esau,[164] Plates 2E, 3D, F and 4C, pp. 106, 107 and 108.)

predominate.[477] With various treatments, primordia which would have developed as male flowers were induced to develop as female spikelets.

In hermaphrodite flowers, the development of the ovary (and the calyx) can be promoted by treatment with NAA, the stamens and corollas being suppressed to some degree.[247] The auxin level can thus affect the relative development of the various floral organs. In monoecious members of the

Fig. 6.32 Diagram of the vascular cylinder of the receptacle of *Ranunculus repens* split lengthwise and spread out in one plane. Traces to floral appendages are labelled as follows: c, carpels; MPB, major pedicel bundle; mpb, minor pedicel bundle; Pe, petal; S, stamens; Se, sepals. (From Tepfer,[500] Fig. 39, p. 563. Originally published by the University of California Press ; reprinted by permission of the Regents of the University of California.)

Cucurbitaceae, e.g. *Cucurbita pepo*, in which staminate flowers are usually formed at the first several nodes, these being later interspersed with pistillate flowers, treatment of the plants with auxin led to a much earlier development of pistillate flowers. Auxin can preferentially stimulate the development of pistillate flowers.[247] In cucumber, treatment with the growth regulator ethrel led to development of female flowers from the outset, no male flowers being formed.[468] Auxin supplied to excised cucumber buds in aseptic culture can also profoundly modify their development.[196]

This and other work on sex expression, which has been reviewed elsewhere,[125, 249, 385] shows that the development of the floral organs can be fairly readily modified hormonally. This is compatible with the view, outlined earlier, that these organs may be expressions of the changing biochemical state of the floral apex, presumably activated by the derepression of a sequence of genes.

Vascular system

If the vascularization of the leafy shoot is complex, that of the compressed floral shoot may be expected to be even more so. Vascular strands are present in association with the various lateral organs. Sepals, petals and carpels often have 3 traces and stamens one; but this varies from species to species. Petals may have a single strand which branches into three. Sepals and petals often show a complex venation system comparable with that of the leaf. The vascular system of the flower of *Ranunculus repens*, the creeping buttercup, spread out in one plane is shown in Fig. 6.32. Considerable variation occurs both in different sectors of the receptacle of the same species[500] and in different species. The vascular system of *Nymphaea*,[370] another ranalian flower, affords a further example of the complexity associated with large numbers of floral parts.

7

Fruits and Seeds

As we have seen in Chapter 6, following fertilization the ovary develops into the fruit and the ovule into the seed. During these processes considerable growth and development are often involved. In particular, many edible fruits (and some seeds) reach remarkable size. It might be supposed that this would impose considerable demands on the original vascular supply to the carpels. Little attention seems to have been given to this aspect of fruit development, although descriptions of the vascular supply to fruits exist. Marked physiological changes occur in ripening fruit, culminating in the rise in respiration known as the *climacteric*, and again the associated anatomical changes appear to have been somewhat neglected. Some discussion of histological changes in the apple is given below.

Fruits and seeds of different species often contrast greatly in structure. Since it is therefore difficult to provide a general description, the structure of a few examples of each will be given. In general, examples of some economic importance have been chosen.

FRUITS

Fruits may be sub-divided into a number of types. For example, they may be dry or fleshy, dehiscent or indehiscent, monocarpellary, apocarpous or syncarpous. In fleshy fruits the *pericarp*, which develops from the ovary wall, is usually substantial; it consists of an outer region, the *epicarp*, a middle parenchymatous region, the *mesocarp*, and an inner *endocarp*, which may be hard and stony. In dry fruits the pericarp often has a high proportion of sclerenchymatous tissue. In so-called false fruits, such as the apple or strawberry, parts of the flower other than the carpels are also involved in the development of the fruit.

In contrast to that of seeds, fruit anatomy is not often used as a taxonomic character. However, a study of some Umbelliferous fruits with the scanning electron microscope indicates that at this level of magnification some microscopic features of the surface may provide useful taxonomic traits.[256]

Fruit ripening usually involves a change in colour, e.g. the change from green to red or yellow in varieties of tomatoes. During this process chloroplasts develop into chromoplasts. Usually the thylakoids of the grana disappear and there is an accumulation of the pigment lycopene.[433] In different genetic lines of tomatoes, differences in the carotenoids are associated with marked differences in plastid ultrastructure.[242] Ripening processes can often be accelerated by treatment with ethylene or with the new growth regulator ethrel, which is similar in its action.[268]

Growth of fruits

As already pointed out, many edible fruits undergo conspicuous growth during ontogeny. This growth may be attributed to the effects of pollination. It is believed that the pollen tubes secrete an enzyme which is capable of converting tryptophane into an active hormone. As the pollen tubes grow down the style a hormone is synthesized which may diffuse into the ovary and induce its growth.[226] Occasionally, mature fruits may develop in this way without fertilization, and parthenocarpic fruits are often induced commercially by treatment with various hormones. These phenomena are more fully discussed elsewhere.[330] Nearly twenty years ago Nitsch[382] showed that unpollinated flowers of tomato grown in sterile culture could be induced to form fruits which increased up to 150 per cent in diameter if treated with an auxin.

Normally, however, growth begins in the fertilized ovules and extends to the placentae and the wall of the ovary. Growth of the nucellus is probably stimulated by auxin associated with pollination.[336] In order to develop, some fruits need only the initial stimulus resulting from pollination, but others, such as the apple, need an additional supply of hormones which apparently come from the young seeds, possibly the developing endosperm and embryo.[226, 336] A further function of this hormone is to prevent abscission of the fruit. The effects of auxin and the formation of the abscission layer in leaves have already been discussed in Chapter 5. The relationships between auxin production, fruit development and fruit drop in apple are shown in Fig. 7.1.

In strawberry, the edible fruits consist largely of receptacular tissue, but this accumulates sugars like a true fleshy fruit, and is also stimulated to develop by auxin. In experiments on developing strawberry fruits it was

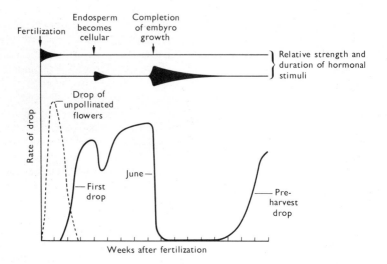

Fig. 7.1 The relationship between endogenous hormonal stimuli and the periods of fruit abscission in apple. Ontogenetic events in the developing fruit are also indicated. (From Luckwill,[336] Fig. 8–3, p. 228. Copyright 1959, The Ronald Press Company, New York.)

shown that if all the achenes which are borne on the receptacle, each of which contains a single ovule, were removed 4 days after pollination or later, growth of the receptacle ceased completely. If some achenes were left, localized growth of the receptacle ensued (Fig. 7.2). If achenes were removed 9 days after pollination and an auxin in lanolin applied to the receptacle, the fruit developed, but if plain lanolin without auxin was applied, it did not (Fig. 7.3). Extraction treatments showed that abundant free auxin was present in the achenes, but none in the receptacle itself.[381] Thus, here also the fruit develops in response to auxin produced in the developing seeds.

The growth of stone fruits is characterized by 3 phases.[530] In stage I the diameter increases rapidly; in stage II growth is relatively slow; and stage III is a period of rapid growth to maturity. Since the embryo grows very considerably during stage II it has been suggested that at this time it may compete successfully with the fleshy tissue of the mesocarp for auxin produced by the endosperm. Application of an auxin to apricot fruits during this phase did increase fruit growth to some degree.[114]

One may ask, what is the effect of these growth regulating substances in anatomical terms? Indeed, our knowledge on this point is scanty. Fruit

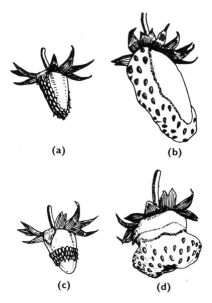

Fig. 7.2 Effect of achenes on fruit growth in strawberry. (a) 3 rows of achenes were left on the young fruit, which developed into a long flat structure (b). (c) 3 horizontal rows were left on the fruit, which developed into a short thick structure (d). In both instances growth of the receptacle was stimulated in the regions where the achenes remained. (From Nitsch,[381] Figs. 3 and 4, p. 214.)

growth comprises both cell division and cell enlargement, the importance of each varying both with species and with stage of development. Presumably, therefore, the growth substances must control the extent and perhaps the planes of cell division and enlargement in the developing fruits. Although further studies, including carefully controlled experimental investigations, are badly needed, the work of Sinnott[469–473] on cucurbit fruits provides a valuable beginning.

Sinnott studied the control of size and shape in the intriguing array of oddly-shaped gourds and squashes belonging to the genera *Lagenaria* and *Cucurbita* in the family Cucurbitaceae. He found[473] that races with disc-shaped and spherical fruits differed by only a single gene, which acted in early stages of development. In tomato fruits (Solanaceae), also, the genes for shape apparently act at a very early stage of development, and affect planes of cell division.[261] Segregation for the length–width indices of cucurbit fruits was sharper than for either length or width, suggesting that the shape itself was the character most directly under gene control. By means of various disc-sphere crosses, Sinnott[473] found that, whereas fruit

9

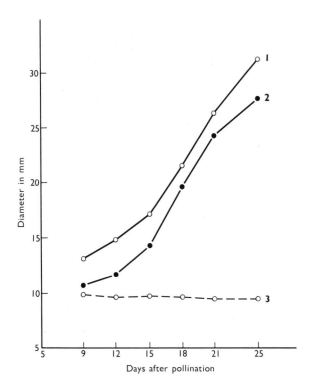

Fig. 7.3 Growth curves of 3 strawberry fruits. **1**. Control, intact fruit. **2**. All achenes were removed on the 9th day after pollination and lanolin paste containing 100 p.p.m. of β-naphthoxyacetic acid (an auxin) applied. **3**. Fruit treated as in **2**, but plain lanolin applied. (From Nitsch,[381] Fig. 5, p. 215.)

size seemed to be determined by a series of multiple genes, the shape difference depended on a single gene inherited independently of size. It is not easy to envisage how a gene controlling shape must act, but some of Sinnott's investigations afford some clues. In long narrow fruits, length increased faster than width but at a constant relative rate. In the much more squat 'bottle' gourd, width increased faster than length. The difference in shape depends on a difference in relative growth rates. However, in 'miniature' and 'giant' races of the bottle gourd the relative growth rate was the same and the shape of the two races at maturity differed because of a difference in *duration* of growth, rather than rate. Since width is increasing faster than length, the fruits grow flatter with increasing size and mature 'giant' fruits differ in shape from mature 'miniature' fruits (Fig. 7.4). Differences in fruit shape depend primarily on the relative numbers of cells along the different axes, and thus genetic control of planes of cell division must be involved.[469] Later analysis of the orientation of mitoses, and thus of the planes of cell division, demonstrated a close relationship between these and the direction of growth of the fruit.[471] Differences in fruit size are usually due to differences in both cell number and cell size. The innermost tissues of the fruits ceased cell division sooner than the more peripheral tissues, but showed a more rapid increase in cell size during the period of mitosis. Formation of secondary cell walls may be one factor which limits cell expansion; forms with the least cell expansion tend to be hard and woody at maturity, whereas fruits with relatively great cell enlargement, like water-melon, tend to remain soft.[470] In large-fruited

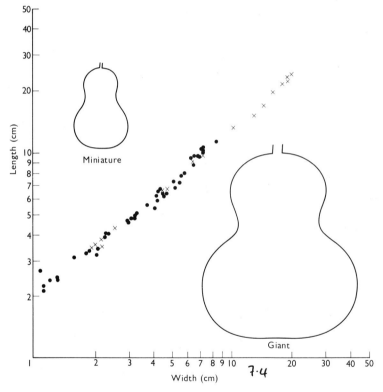

Fig. 7.4 Relative growth of length and width in 'bottle' gourds, plotted logarithmically. The points fall along a straight line. Width increases faster than length, but the relative rate of growth is the same in the 'miniature' (dots) and 'giant' (crosses) races. (From Sinnott,[469] Fig. 2, p. 250. Copyright by the University of Chicago Press.)

races of cucurbits each portion of the growth cycle is of longer duration than in small-fruited ones, and it is suggested that the genes controlling fruit size may control the amount of substances necessary for fruit growth.[472] Clearly, there is abundant scope for study of the effects of various hormones on the control of cell division and cell expansion in developing fruits.

An interesting example of elongation of the fruit stalk, resulting from the activity of an intercalary meristem, is seen in the gynophore of the peanut, *Arachis hypogaea* (see Chapter 3).[269] The intercalary meristem is situated just below the fruit (Fig. 7.5a), and its activity leads to the development of an elongated structure, the gynophore (Fig. 7.5b), the downward growth of which results in eventual burial of the developing fruit below the surface of the soil. Recent work indicates that gynophores at the upper nodes of the plant, which do not normally set pods, could be induced to do so and to elongate by causing the gynophores to grow through cheese-cloth soaked in GA.[13] Thus, GA is apparently frequently involved in the stimulation of intercalary growth of this type (see also Chapter 3).

Fruit abscission

Fruit drop is an important source of loss in commercial orchards, and thus fruit abscission assumes some economic importance. The general process of abscission has been discussed in Chapter 5. Some thirty years ago Barnell[33] described the formation of the abscission layer in the fruit stalk of mango, *Mangifera indica*, and avocado, *Persea americana*. It is interesting to note that here, as in abscising leaves, tyloses are abundant in the xylem. In avocado, separation occurs at the middle lamella of the cells. Barnell[33] points out that fruit fall occurs as a normal phase in the progressive softening of the fruit tissue during later stages of maturation. The phase of ripening involving the dissolution of the middle lamella is responsible for the shedding of the fruit. The fruit stalk, which remains on the tree, is then shed by the formation of an abscission layer, comparable with that in a leaf.

Fruit abscission is further discussed elsewhere.[330] More recent work on cotton fruits has shown that abscission-accelerating substances, abscisic acid and GA, were very high in young fruit, and subsequently declined. The maximum activity of these substances coincided with the period of maximum shedding of the immature cotton bolls.[479] Thus, these hormones evidently have an important effect on the undesirable fruit drop in early stages of development.

Structure and growth of the sour cherry

As already mentioned, the cherry and other stone fruits exhibit 3 fairly well defined stages of growth. Most of the increase in size during growth is

Fig. 7.5 Developing gynophore of Starr peanut. (a) Longitudinal section of the tip of the gynophore. Two ovules (o) are contained within the developing fruit; a conspicuous intercalary meristem (im) is present just at the base of the fruit. × 75. (b) Excised node with two flowers and a developing gynophore, growing downwards. × 1·5.

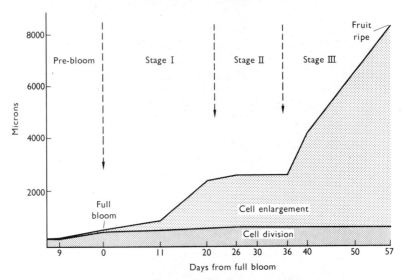

Fig. 7.6 Increase in thickness of the pericarp of the sour cherry (*Prunus cerasus*) from the pre-bloom stage to fruit ripening, showing the relative amount of growth attributable to cell division and cell enlargement. (From Tukey and Young,[530] Fig. 8, p. 745. Copyright 1939 by the University of Chicago. All rights reserved.)

a consequence of cell enlargement (Fig. 7.6). Mitoses were not found in the mesocarp region after the 10th day from full bloom, and divisions in the endocarp had ceased by the middle of stage II.[530] The structure of the cherry fruit 40 days after full bloom is shown in Fig. 7.7. The description which follows is taken from the work of Tukey and Young.[530]

The skin of the cherry consists of the epidermal and hypodermal layers, the hypodermal layer being collenchymatous. The fleshy pericarp, or **mesocarp**, comprises an inner layer of thin-walled parenchyma and a more peripheral region of larger parenchyma. The stony pericarp, or **endocarp**, also consists of inner and outer layers. At an early stage of fruit development the future stony region, which develops from a distinct group of cells, consists of smaller cells than the fleshy part. As a result of rapid division prior to full bloom, the inner region becomes about 4 cells thick and the outer part 10 cells thick. It increases approximately 5-fold during stage I, and the cell walls thicken and harden during stage II. The hardening of the cell walls progresses basipetally. Most of the cells are sclerenchymatous by the time the fruit ripens.

The increase in width of the mesocarp region occurs principally in the pre-bloom phase and the early part of stage I. After the end of stage I there

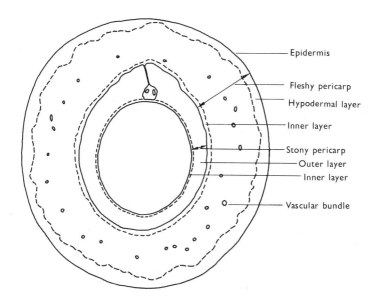

Epidermis

Fleshy pericarp

Hypodermal layer

Inner layer

Stony pericarp
Outer layer
Inner layer

Vascular bundle

Fig. 7.7 T.S. fruit of the sour cherry 40 days after full bloom, showing the various structures present. (From Tukey and Young,[530] Fig. 2H, p. 729. Copyright 1939 by the University of Chicago.)

is no further cell division, except perhaps in the region nearest the endo-carp. Earlier, cell division and also cell enlargement lead to a 7-fold increase in thickness. Intercellular spaces are evident in the thin-walled fleshy region. Vascular bundles are present in the mesocarp, forming a ring. During stage III the cells increase markedly in size, and intercellular spaces are much less conspicuous.

Stomata are present in the epidermis, or **epicarp**, which is covered by a cuticle. Cell division occurs during stage I, cell enlargement predominantly during stage III (Fig. 7.6).

Structure and growth of the apple

The apple fruit is sometimes considered to be composed partly of a fleshy development of the receptacle, or alternatively of the fused and enlarged basal parts of the floral tube. In developing fruits a 'core line', consisting of smaller cells and separating the outer fleshy part of the fruit from the core, is discernible (Fig. 7.8); this is interpreted by adherents of the fleshy floral cup theory as the line of fusion between the ovary and the other, swollen parts, or, by those supporting the receptacular theory, as a 'cambial region' between the cortex and pith.

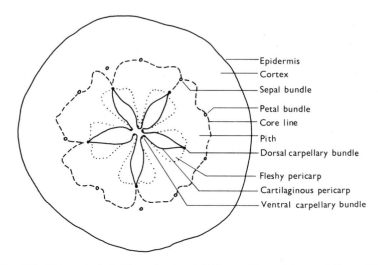

Fig. 7.8   T.S. apple fruit 14 weeks after full bloom. (From Tukey and Young,[531] Fig. 14, p. 10. Copyright 1942 by the University of Chicago. All rights reserved.)

Considerable cell division follows fertilization of the apple flower, and a rapid increase in the diameter of the fruit occurs 5–6 days after pollination.[342] These cell divisions occur throughout the developing fruit. The duration of cell division and cell enlargement differs in different regions of the fruit. The growth curve for the whole fruit is shown in Fig. 7.9. It does not exhibit stages comparable to those of stone fruits. The inner cartilaginous endocarp develops rapidly fairly early in the season, and reaches its maximum size before any other region. The fleshy mesocarp continues growth for a further two months. In some varieties the edges of the carpels do not completely unite, giving an 'open core'; in other varieties union is complete, and there is a 'closed core'.[341]

The endocarp is formed from the layer of cells just below the inner epidermis of the carpel. These 5 or 6 layers of cells become thick-walled and sclerenchymatous. As a result of the continued growth of the surrounding tissues, the endocarp may become split.[531]

The fleshy mesocarp consists of thin-walled parenchyma with intercellular spaces, and considerable cell division and cell enlargement take place in this region. The later growth of the fruit is primarily by cell enlargement which also ceases about 150 days after pollination.[376] Vascular bundles traverse the mesocarp. The inner bundles supply the placentae, ovules and core.[504] The mesocarp consists of fairly uniform parenchyma,

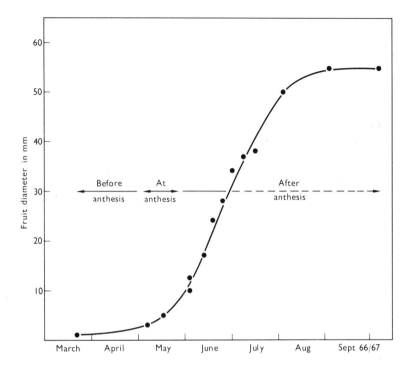

Fig. 7.9 Seasonal growth in diameter of the apple fruit. Horizontal lines indicate various developmental stages. (After De Vries,[140] Fig. 1, p. 231.)

and these cells must be the principal ones involved in the physiological changes associated with ripening, including the rapid rise in respiration known as the climacteric, which has been the subject of much physiological work. The structure of the cell wall in apples from trees given different fertilizing treatments, and having different keeping qualities, has recently been compared.[376] Of those studied, fruits from trees fertilized with sulphate of ammonia, superphosphate and sulphate of potash (NPK apples) had the shortest storage life, those given P and K and grown in grass sward (PKS apples) the longest. Under the latter conditions the supply of nitrates would be poor. The PKS apples had larger cells, but also nearly twice as much wall material per cell, and more material per unit area of wall, than the NPK apples during a certain period. The wall material reached a maximum at 110–120 days after pollination. There seemed to be a difference in the basic structure of the cellulose during the phase of most rapid cell growth as a result of the different cultural treatments. However, it is considered unlikely that the superior keeping qualities of the PKS apples are attributable directly to the wall structure; more probably low levels of nitrogen during growth result in the development of characteristics in both the wall and the protoplast which contribute to maintaining cells in a healthy condition for a long period.[376] It will be interesting to pursue the precise nature of these desirable features.

The skin or epicarp of the apple consists of the epidermis and hypodermis. In early stages of growth numerous hairs are present.[504, 531] Cell division continues for some time. Stomata are present during early growth, but apparently are later replaced by lenticels.[504] A cuticle is present overlying the epidermal cells; its structure does not change during increase in thickness.[140] In apples with russeted skins, such as the variety Cox's Orange Pippin, cork is formed.[141, 504] Suberized cells may occur in isolated regions of the fruit, either naturally or in response to wounding; in either case their structure is similar.[141]

Structure and growth of citrus fruits

Citrus fruits have a rather unusual structure in that the edible portion consists of *juice sacs*, outgrowths from the endocarp into the locules. The following description is based on the work of Schneider.[453] During development of the fruit the approximately 10 carpels fuse in such a way that the wings of the 'lamina' of the carpels become the septa which enclose the locules. Placentation is axile. The septa consist of two layers of endocarp-like tissue with mesocarp-like cells between. In some citrus fruits the mesocarp-like tissue is easily torn, and the fruit can be separated into segments. Each segment of an orange, for example, consists of a locular membrane (endocarp-like tissue) enclosing juice sacs and seeds.

The endocarp is the inner part of the pericarp and part of the locular membrane. It is made up of the inner epidermis of the carpels and several layers of compact parenchyma. The mesocarp or *albedo*, which is white, is made up of branching cells with many intercellular spaces and resembles the spongy mesophyll of leaves. Vascular bundles ramify through the mesocarp. The coloured epicarp or *flavedo* includes the epidermis with cuticle and adjacent parenchyma. Oil glands, responsible for the aroma of the fruit, and idioblasts containing crystals are present in the epicarp (Fig. 7.10). In immature, green fruits chloroplasts are present in the parenchymatous cells; when the fruit ripens these develop into chromoplasts. As in tomato fruits, the grana-fretwork system of the plastids breaks down and there is a loss of chlorophyll and increase of carotenoids, which give the ripe fruit its characteristic colour. The peel of citrus fruits consists of the epicarp and most of the mesocarp.[453]

The onset of ripening and senescence in citrus fruits can be delayed by

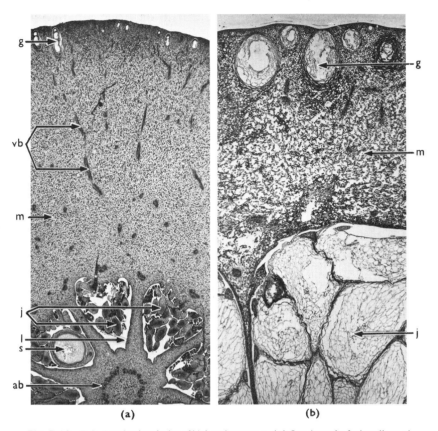

Fig. 7.10 T.S. developing fruits of Valencia orange. (**a**) Section of a fruit collected in July. The pericarp has enlarged greatly by cell division, but the locules (l) are still relatively small. Juice sacs (j) are present, and one seed (s). Oil glands (g) occur in the outer region of the pericarp (epicarp). The mesocarp (m) or albedo is traversed by vascular bundles (vb). Axial bundles (ab) are present centrally.   × 12. (**b**) Section of a mature fruit, showing the pericarp and parts of two locules. The locules and juice sacs are now relatively much larger. Intercellular spaces are evident in the mesocarp region.   × 12. (From Schneider,[453] Figs. 1–22D and 1–30A, pp. 67 and 83.)

treatment with gibberellic acid (GA). In the navel orange, the fruit may not be harvested until as much as 8 months after softening of the rind or peel begins. Softening is a result of enlargement and vacuolation of the cells of the flavedo and albedo; many intercellular spaces form and the cell walls weaken and break, especially in the albedo. After treatment with GA, however, these cells remain intact and the tissue is more compact.[107]

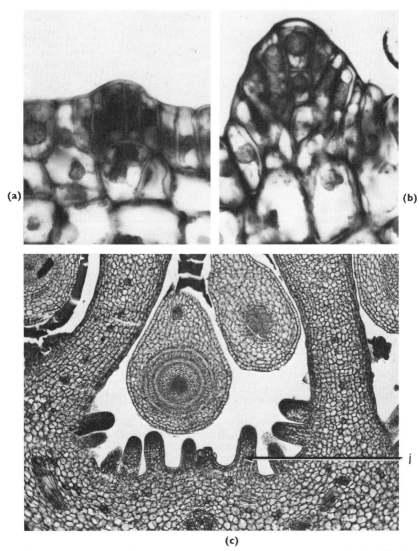

Fig. 7.11 Development of juice sacs in the Valencia orange. (**a**) Anticlinal divisions have occurred in the epidermal layer and a periclinal division in the sub-epidermal layer, giving rise to a juice sac primordium.   × 1100. (**b**) Divisions have occurred in the epidermal and sub-epidermal cells of the juice sac primordium. × 1300. (**c**) Section of a locule in which the juice sacs (j) have elongated and show meristematic growth at the tip.   × 100. (From Schneider,[453] Fig. 1–24A, E and H, p. 71.)

The juice sacs are stalked outgrowths from the endocarp. In the Valencia orange, juice sacs originate as dome-shaped projections from the endocarp into the locules (Fig. 7.11); they are formed from the time of anthesis until the style falls. Epidermal and usually subepidermal cells participate in their formation. The tip of the juice sac remains meristematic and continues growth. Cell enlargement begins in the centre of the sac and progresses outwards. At maturity, the juice sacs consist of an epidermis of elongated cells with a waxy cuticle, enclosing large vacuolated cells which contain the juice. A study of developing juice sacs with the electron microscope would be likely to prove interesting.

In Valencia orange, growth can be divided into three stages. In stage I the tissues of the pericarp undergo cell division for several months; most of the size change of the fruit is attributable to growth of the mesocarp, the locules enlarging little. During the six months of stage II the pericarp grows by cell enlargement and also shows differentiation, and the locules increase in size. In stage III, growth slows down and the fruit changes colour and matures.[453]

Recently it has been found possible to culture lemon tissue *in vitro*,[373] a technique which has some important potentialities.

Structure and growth of the banana

The banana has been selected and bred over a long period of time for succulence of the fruit and a lack of seeds. The development of parthenocarpic and seed-bearing varieties of *Musa acuminata* has been most recently described by Mohan Ram, Ram and Steward.[364] As in other seedless varieties of fruits, it is believed that parthenocarpic bananas may contain a greater amount of endogenous auxin than the unpollinated fruits of seeded varieties.

As is evident from Fig. 7.12a and c, at the time of emergence the ovaries of parthenocarpic and seeded bananas are similar, but later the parthenocarpic fruits are characterized by the growth of a soft, fleshy *pulp*, whereas in the seeded fruits the seeds themselves grow and contribute to the increase in size (Fig. 7.12b and d).

The epidermis has a cuticle on the outer surface. Several layers of hypodermal parenchyma occur. Internal to these is a broad region of parenchyma with scattered vascular bundles. Towards the interior of the fruit is a region of parenchyma with well developed air spaces, and internal to this region is a zone with vascular bundles surrounded by laticifers, running at right angles to the long axis of the fruit. The interior of the pericarp consists of 5–7 layers of parenchyma which contains the pulp-initiating cells and is bounded by the inner epidermis.

The septa consist of central parenchyma, parallel vascular bundles, with hypodermal pulp-initiating cells and epidermis on either side.

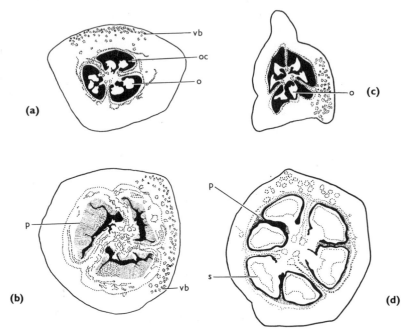

Fig. 7.12 T.S. fruits of developing parthenocarpic and seedbearing varieties of banana. (**a**) Parthenocarpic fruit at the time of emergence of the inflorescence. (**b**) Parthenocarpic fruit 8 weeks after emergence, showing the pulp (p) invading the ovarian cavity (oc) or locule. (**c**) Seeded fruit at the time of emergence. (**d**) Seeded fruit 8 weeks after emergence. Very little pulp is present around the locules, which are occupied by the enlarging seeds. o, ovule; s, seed; vb, vascular bundle. (From Mohan Ram, Ram and Steward,[364] Figs. 44, 47, 70 and 73, Plates 6 and 10.)

In parthenocarpic bananas cell divisions occur in the pulp cells by two weeks after emergence of the inflorescence. Cell expansion occurs in the pericarp. The ovules degenerate, and the septum expands into the locule, partly as a result of the activity of the pulp-initiating region. Cell division in the precursors of the pulp continues until about 4 weeks after emergence, and is succeeded by cell enlargement. Activity of these cells is not uniform, and the pulp often has an irregular outline. By continued growth the pulp almost occludes the locules (Fig. 7.12b). Starch is deposited in the pulp from about 4 weeks after emergence and begins to disappear about 8 weeks later.

In the banana fruit two kinds of tannin-containing elements occur: laticifers in the pulp and skin or peel, and scattered tanniniferous parenchyma cells in the skin.[34] Unripe bananas have an astringent taste which is attributed to the tannins; during ripening these astringent properties are lost. The amount of tannins decreases considerably in both skin and pulp of ripe bananas; at this time the latex also dries up and tannins are no longer detectable in the dried latex.[34] Since the astringent properties are of course an undesirable feature of the fruit, it would be interesting to discover what factors control the secretion of the tannin. This is just one way in which anatomists and physiologists might assist fruit growers, which on the whole the former, at least, seem to have been rather slow to do.

The banana peel separates from the pulp at the region of parenchyma with conspicuous air spaces, which have meanwhile enlarged considerably.

In the seeded variety studied,[364] enlargement of the ovules, expansion of the septa and increase in size of the ovary occurred in about a week. After 4 weeks an increase in the number of pulp cells is evident, but the seeds nearly fill the locules. In later stages of development the pulp tissue remains relatively meagre (compare Fig. 7.12b and d). The seeds become very hard.

It has been pointed out that in edible bananas, in contrast to most other fruits, it is the failure, rather than the success, of fertilization that allows the ovary to grow and induces growth of the pulp. The presence of fertile seeds suppresses the development of this tissue.[364] Complex hormonal controlling mechanisms must be involved, and offer a promising and economically important field for further investigation.

Structure and growth of the legume

In contrast to the fleshy fruits described above, the legume, of which the pea pod is a common example, is largely composed of sclerenchymatous tissue. Legumes are generally characterized by an outer epidermis with underlying parenchymatous tissue, then sclerenchymatous tissue. Usually a few layers of parenchyma are interposed between this and the inner epidermis. Vascular bundles are present in the outer region of parenchyma.[174]

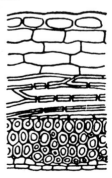

Fig. 7.13 T.S. part of a pod of *Astragalus hamosus*. Note thick-walled sclerenchyma. (From Fahn and Zohary,[174] Fig. 11, p. 102.)

Some pods are dehiscent, others are indehiscent, and there is usually a correlation between the anatomy of the pod and the degree of dehiscence. For active dehiscence to occur, two factors are considered to be necessary; the cross orientation of the sclerenchymatous elements or of their cellulose micelles, and the presence of a separation tissue in the region of the suture extending from one epidermis to the other.[174] Fig. 7.13 shows part of a section of the pod of *Astragalus hamosus*, in which the long axis of the cells comprising the inner layer of sclerenchyma runs parallel to the longitudinal axis of the pod, while that of the outer layer runs transverse to it.

SEEDS

During the development of the seed from the ovule, the integument, or integuments, develop into the seed coat or **testa**. The fertilized egg or zygote develops into the **embryo** (see Chapter 8), and the triple fusion nucleus, or primary endosperm nucleus, divides to give rise to the **endosperm**. Exalbuminous seeds have little or no endosperm; in such seeds storage of food materials is usually restricted to cells of the cotyledons.

In some species, a brightly coloured structure, the **aril**, develops as an outgrowth of the funicle. A well-known example is the spice, mace, which is the aril of the seed of nutmeg, *Myristica fragrans*.

The seed coat or testa is important for a number of reasons. It may affect the dormancy of the seeds; such effects may be attributable to interference with oxygen uptake. Or the testa may offer mechanical resistance to the growth of the radicle. The seed of *Xanthium*, the cocklebur, contains two water-soluble growth inhibitors. These are rapidly leached out of the embryo itself, but the testa is impermeable and leaching therefore does not occur from intact seeds.[567] Hardness of the testa may also offer protection, and seeds which are viable for long periods of time, such as those of the sacred lotus, *Nelumbium*, often have very hard and relatively impermeable seed coats.

Working on the control of dormancy in rice, Roberts[430] found that treatment with respiratory inhibitors stimulated the breaking of dormancy both in rice and in several other species. Partial removal of the testa also broke dormancy. Histochemical studies showed that both the aleurone layer at the periphery of the endosperm and the testa had a high activity of enzymes involved in oxidation-reduction reactions. Thus it is believed that the outer layers of the seed may not only constitute a physical barrier impeding access of oxygen to the embryo, but these layers may actually compete for oxygen with the embryo.[431] A very specific physiological effect of certain layers of tissue is thus demonstrated, as it is also in the induction of α-amylase synthesis in the aleurone layer, discussed below. In charlock, *Sinapis arvensis*, dormancy is considered to be maintained by the action of a specific growth inhibitor which is produced at low oxygen concentration in the interior of the seed.[161, 162] Again, the testa is thought to impede the entry of oxygen, perhaps because of the mucilages and phenols which occur in it.[160] It is interesting to note that charlock seeds also have an aleurone layer.

Seed structure is of great importance in taxonomy. In the family Leguminosae seeds have very characteristic features; the testa has an external palisade tissue, which develops from the outer epidermis of the outer integument, and also hour-glass-shaped cells, which usually develop from the hypodermis of the outer integument (Fig. 7.14). Even a fragment

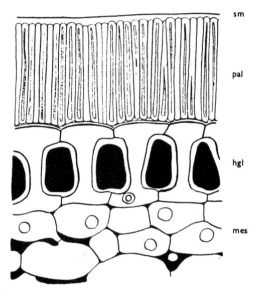

Fig. 7.14   T.S. outer part of the testa of *Mucuna utilis*, showing the thick-walled palisade cells and the hour-glass-shaped hypodermal cells. hgl, hour-glass cells; mes, mesophyll of external integument; pal, palisade; sm, mucilage-stratum. (From Corner,[110] Fig. 1A, p. 118.)

of seed having these characters can be identified as leguminous.[110] By means of various surface features, many seeds of economic importance can be identified to the species, or sometimes even the variety.[540] Very small fragments are often sufficient to allow identification, a fact which is important in forensic medicine and in controlling the quality and purity of foods and drugs, since adulterants can be detected relatively easily.

As variation in seed coat structure is so great that it can be used in classification, it is evident that once again no effective generalizations can be made. Accordingly, a few examples will be described, in order to illustrate something of the range of structure that occurs.

Structure of the seed coat

Brassica (*Cruciferae*)

In the seeds of members of the family Cruciferae, to which *Brassica* belongs, the outer epidermis of the testa is often mucilaginous. This is well seen by soaking seeds of cress, *Lepidium*, when the mucilage becomes readily apparent. So-called palisade cells, elongated cells with thickened radial and inner tangential walls, are also often present in members of this family. A pigment layer, representing the remains of the inner integument, may also occur.[540]

In seeds of *Brassica* spp. the epidermal cells are elongated as seen in transverse section. The outer wall is convex and mucilage is often associated with it, occupying the entire lumen (Fig. 7.15). A distinctive sub-epidermal layer may be evident, or not (compare Fig. 7.15a and b). The

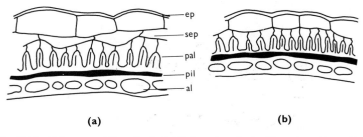

(a)   (b)

Fig. 7.15 T.S. testa of *Brassica integrifolia* var. *carinata*. The sub-epidermal layer may be present (**a**) or obliterated (**b**). ep, epidermis; sep, sub-epidermal layer; pal, palisade; pil, pigment layer; al, aleurone layer. (From Vaughan,[538] Figs. 1 and 2, p. 365.)

palisade cells have the inner walls and part of the radial walls thickened and slightly lignified.[538] In distinguishing seeds of different species of *Brassica*, the size and shape of the epidermal cells may be important. Most rape seeds have a dark pigment in both the palisade and pigment layers, but some varieties do not.[539]

Gossypium (*Malvaceae*)

The epidermal cells of the ovule of cotton, *Gossypium*, elongate into long unicellular hairs (see Part 1,[127] Fig. 7.7). Thus the outer epidermis of the mature testa bears these hairs, which constitute commercial cotton (Fig.

7.16). There is a sub-epidermal zone of several layers of cells, some of which are pigmented, a layer of extremely elongated palisade cells, and an inner region of several layers of cells.[540] The palisade cells have very narrow lumina, and each cell shows a 'light line'. This is a band which runs transversely to the long axis of the cells, and often occurs in palisade cells. Light refraction differs in this region, and it is believed that the cell wall is

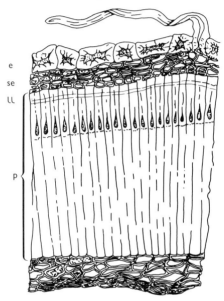

Fig. 7.16 T.S. testa and cotyledon of *Gossypium*, cotton. e, epidermis; se, sub-epidermal layer; ll, 'light line'; p, palisade layer. × 270. (From Vaughan,[540] Fig. 1A, p. 252. Published by permission of The Linnean Society of London.)

chemically and physically modified in some way. Other members of the Malvaceae show a somewhat similar palisade layer. The testa of *Ceiba pentandra*, the source of Kapok, which belongs to the Bombacaceae, sometimes considered a related family, is very similar.[540]

Cucurbita (*Cucurbitaceae*)

The cells of the outer epidermis of the seed coat of *Cucurbita* are extremely elongated, and have branching ribs on the radial walls (Fig. 7.17). Several layers of small pitted cells lie beneath the epidermis. A single layer of large stone cells intervenes between this and a region of spongy parenchyma cells, some of which have walls with reticulate thickenings. A similar pattern is found in other members of the family.[540]

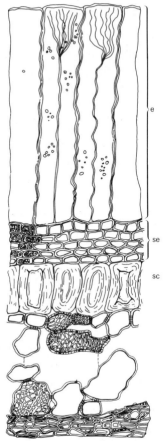

Fig. 7.17   T.S. testa of seed of *Cucurbita*. e, epidermis; se, sub-epidermal layer; sc, stone cells. × 200. (From Vaughan,[540] Fig. 1H, p. 252. Published by permission of The Linnean Society of London.)

Phaseolus (*Leguminosae*)

As already mentioned, the outer epidermis of leguminous seeds is composed of palisade cells, and the hypodermal layer of so-called hour-glass cells. These layers occur in the seeds of *Phaseolus* (Fig. 7.18). Below them is a region of thin-walled parenchymatous tissue with vascular bundles. The outer part of this tissue has well developed intercellular spaces and somewhat resembles spongy mesophyll. The region of the **hilum** has a very specialized organization.[110] The attachment of the funicle is expanded into a disc-shaped structure which fits into the depression of the hilum. The outer layer of cells in the head of the funicle con-

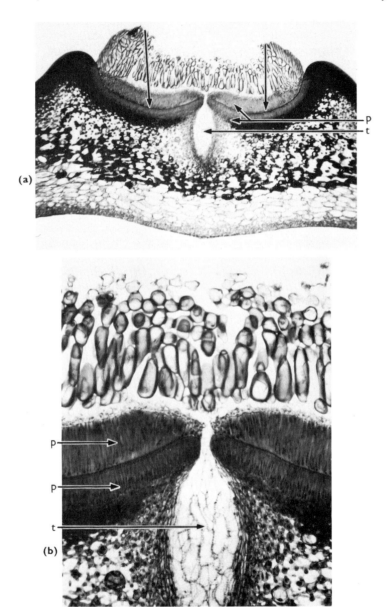

Fig. 7.18 T.S. seed coat of French bean, *Phaseolus vulgaris*, in the region of the hilum. Above the arrow heads is tissue of the funiculus. Palisade tissue (p) consists of macroclereids, t, tracheids. (**a**) ×150. (**b**) ×300.

stitutes a palisade layer, which adheres to the palisade layer of the testa (Fig. 7.18). Both palisade layers are interrupted centrally by a narrow groove, which serves as an air passage in the ripe seed. This groove leads to a group of tracheids, the so-called 'tracheid bar'. Tissue with conspicuous air spaces is present on either side of the group of tracheids. The structure of the bean hilum is described in more detail by Corner.[110]

Experiments [263a] have indicated that the testa of leguminous seeds is impermeable to water except at the hilum. This region apparently functions as a hygroscopic valve, opening under conditions of low humidity and closing when humidity is high, thus retaining the dry condition of the seed during seed ripening. The light line seems to indicate the location of the impermeable layer. Moisture could be absorbed under conditions of gradually increasing relative humidity, in which the hilar fissure remained open.

Endosperm

The development of the endosperm and its relationships with the embryo have been fully reviewed by Brink and Cooper.[64]

The cells of the endosperm often contain reserve food materials, in the form of starch grains or protein bodies. In some species, e.g. date, persimmon, the cell walls may be extremely thick and themselves serve as a food reserve. Conspicuous primary pit fields or pits may be present in the walls (see Part 1,[127] Fig. 4.1). Where perisperm is present in the seed, the cell walls may also be extremely thick and pitted and may be resorbed during growth of the embryo and germination.[18] In some seeds, spherosomes may be present. Various enzymes are localized in these bodies.[491] In tobacco endosperm these organelles accumulate and mobilize reserve lipids. Organelles known as glyoxysomes, in which the enzymes of the glyoxylic acid cycle are localized, also occur in seeds.

Where endosperm is scant or lacking, reserve food materials, such as starch and protein bodies, may be stored in the cells of the cotyledon (Fig. 7.19).

In cereals, the reserves of the endosperm include hemicelluloses of the cell walls, starch and protein. Hydrolysis of these reserves is carried out sequentially by various enzymes.[343] Protein bodies, or aleurone grains, may occur in particular layers of the endosperm, as in the aleurone layer of the grains of cereals. In barley, these cells respond to the hormone gibberellic acid (GA), produced by the embryo, by secreting the enzyme α-amylase.[536] This digests the starch present in the rest of the endosperm, rendering the stored carbohydrate available to the growing embryo. These important physiological interactions between the embryo and regions of the endosperm are more fully discussed in Part 1,[127] Chapter 3, and in another volume in this series.[497] Recently the fine structure of the aleurone cells of barley has been studied, both in dry and imbibed seeds and after treatment with GA. Aleurone grains and spherosomes both occur in the

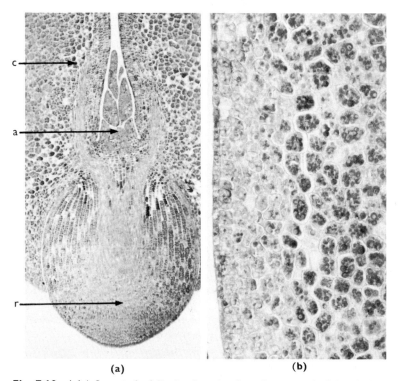

(a)   (b)

Fig. 7.19 (a) L.S. part of a fully developed embryo from a seed of *Acacia longifolia*. a, shoot apex; c, cotyledon; r, radicle.   × 45. (b) Part of a cotyledon enlarged, showing protein bodies in the cells.   × 150. (Slide by courtesy of Mr. H. Chiu.)

aleurone cells of dry and imbibed seeds, and there is abundant rough endoplasmic reticulum (ER). Polyribosomes are also present in the cells.[290] Following GA treatment there is a lag phase during which no significant enzyme synthesis occurs. However, during this time various cytological changes take place. For example, the aleurone grains swell considerably and the amount of rough ER increases. Later the ER becomes distended, reaching a maximum 14 hours after treatment with GA. Probably this represents the period of maximum synthesis of the enzyme.[291] The ER then proliferates vesicles. The amount of ER decreases from 19–22 hours after treatment.[292] The biochemical changes in the cells can thus be related to associated structural changes.

The development of the embryo and its relationships with the surrounding tissues of the seed are considered in the following chapter.

8

Embryos

The structure and development of embryos in the plant kingdom is a vast topic, and one beyond the scope of this book. For detailed descriptions of embryogenesis in different species the reader is referred to the classical works of Souèges,[490] Johansen,[285] Maheshwari,[344] Wardlaw[553] and Davis.[134] For many years workers on plant embryos were preoccupied primarily with the sequence and planes of cell division during development. More recently, however, attempts have been made to discover the nutritional requirements of embryos empirically by removing developing embryos at various stages and growing them in aseptic culture. The composition of plant endosperm is often complex and embryos *in situ* are usually well supplied with nutrients, hormones and other substances. The liquid endosperm of the coconut, coconut milk or coconut water, contains a complex array of inorganic ions, amino acids, nitrogenous compounds, organic acids, vitamins, sugars and growth substances. For that reason, it is often used as a constituent of culture media, for growing both embryos and other plant parts. Endosperm of other species is also known to contain auxin. Various experiments involving treatments of developing embryos with radiation, or chemicals, including hormones, have also been carried out and have yielded much interesting information. This work has been well summarized elsewhere.[418, 553, 559]

Fig. 8.1 L.S. parts of ovules of *Capsella bursa-pastoris*, shepherd's purse, with developing embryos. The suspensor (s) with a large basal cell (b) is evident in each. **(a)** 8-celled globular stage (octant) of the embryo (e).   × 300. **(b)** Later globular stage in which the protoderm (immature epidermis) has been delimited.   × 300. **(c)** Heart-shaped stage, in which two cotyledons (c) have been formed.   × 150. **(d)** Late torpedo stage, with bending cotyledons. p, procambium.   × 150.

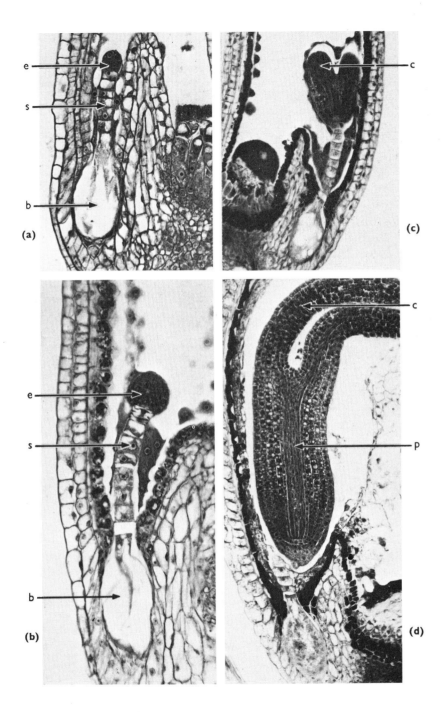

Most embryos pass through a spherical or **globular** phase of development. In dicotyledons the organization of the two cotyledons gives the embryo a **heart-shaped** configuration, and later elongation of the embryo axis leads to a stage described as **torpedo-shaped** (Fig. 8.1). In some species or horticultural varieties, embryos may develop not only from the zygote but also from other parts of the ovule. Nucellar embryos are common in citrus fruits, for example. Such embryos are sometimes called **adventive**.

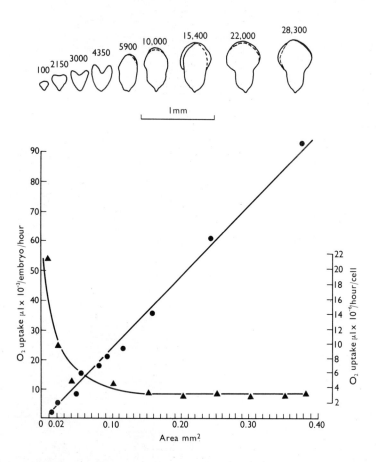

Fig. 8.2 Oxygen uptake in embryos of cotton, *Gossypium hirsutum*. Average uptake per embryo per hour (left ordinate) and average uptake per cell per hour (right ordinate). Developmental stages of embryos, and cell numbers, are shown at the top. (From Forman and Jensen,[181] Fig. 4, p. 767.)

Considerable information on the physiology and growth of animal embryos exists, but it is much more difficult to work with isolated plant embryos and less information is available. In one study with cotton embryos Cartesian diver microgasometers were used to measure respiration of embryos at different stages of development.[181] It was found that the average amount of oxygen consumed per hour increased in a linear fashion as a function of embryo size, which was used as a measure of stage of development rather than age, since the rate of growth varied with the season. If oxygen uptake is considered on a per cell basis, however, it is seen to be high initially, in the globular stage, but falls rapidly in the middle of the heart-shaped stage and levels off to a constant value in the early torpedo stage (Fig. 8.2). The distribution of activity of the respiratory enzyme succinic dehydrogenase, determined histochemically, was associated with areas of active growth and differentiation (Fig. 8.3). In the globular embryo, activity was evenly distributed in the embryo, but rather higher

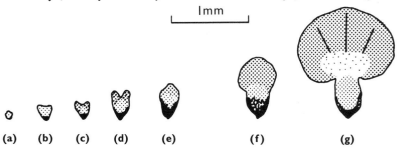

Fig. 8.3 Diagram showing the distribution of activity of the enzyme succinic dehydrogenase in early stages of development ((a)–(g)) of cotton embryos. (From Forman and Jensen,[181] Fig. 10, p. 768.)

in the suspensor region; later, activity was associated with the developing cotyledons. At first, little activity was associated with the apical meristem. Rather similar results were obtained in *Capsella*, in which the distribution of acid phosphatase and cytochrome oxidase as well as that of succinic dehydrogenase was investigated.[418]

In order to complete the description of the structure of all parts of the plant, embryogenesis in a selected dicotyledon and monocotyledon will be described, with reference to work on other species where appropriate. It goes without saying that the zygote, endowed with all the genetic information necessary for the development of the whole organism, is a cell of great interest. Clearly it is very important to understand the structure of the embryo, from which all parts of the mature plant are derived during ontogeny. However, it should be realized that a description of only a few species can give very little idea of the complexity of embryo development in the plant kingdom as a whole. Recent work with the electron microscope, and also histochemical studies of developing embryos, have contributed materially to our understanding of fertilization and embryogenesis, especially the early stages, and these findings are discussed below.

Fertilization

Although the major events in the process of double fertilization—the fusion of one sperm nucleus with the egg nucleus, and the other with the fusion or secondary nucleus—have been known for many years, many unsolved mysteries remain. Recent studies of embryogenesis in cotton, *Gossypium hirsutum*, by Jensen and co-workers[278–283] have contributed significantly to a fuller understanding of the events involved, but clearly there is much scope for further work.

After the pollen grain germinates on the stigma of the cotton flower, the two sperm cells move down the pollen tube together; they are not separated by cytoplasm of the pollen tube. They are distinct, separate cells, each surrounded by a plasma membrane. However, they are extremely reduced cells, being little more than a container for the nucleus.[282] The two cells are ultrastructurally identical, although they will fulfil different functions. It has been suggested that the sperm of many species may lack plastids, which would be of importance in the matter of plastid inheritance. Fertilization occurs 12–24 hours after pollination.[278]

Before the pollen tube reaches the nucellus, a column of cells in the nucellus degenerates; the pollen tube grows between the walls of these cells, just as it has already grown between the walls of the cells of the transmitting tissue in the style. The stimulus that results in the degeneration of these nucellar cells evidently comes, not from the pollen tube, but from the embryo sac or the nucellus itself.[280] The pollen tube grows into one of the two synergids, and the sperm and part of the cytoplasm enters the synergid. The sperm nucleus then enters the egg from the synergid, moves through the cytoplasm of the egg cell and becomes appressed to one end of the egg nucleus. It is not known how the sperm moves; further work will be necessary on this point. Meanwhile, the other sperm nucleus reaches the two polar nuclei, which began to fuse before pollination but have not yet completed the process. The triploid primary endosperm nucleus is formed before the zygote nucleus. Histochemical tests show that the cytoplasm of the central cell, containing the polar nuclei, is high in RNA and protein. It is apparently a more active cell than the egg cell, which is highly vacuolate and less rich in RNA and protein.[281] Following fertilization, the primary endosperm nucleus divides to form the endosperm and the zygote develops into the embryo. The egg, zygote and young embryo of cotton contain a unique kind of endoplasmic reticulum (ER), which contains tubes.[279] Its function is not yet understood.

Embryogenesis in *Capsella*

The embryo of *Capsella bursa-pastoris*, the shepherd's purse, has long been studied by embryologists, and forms the subject of a classical study by Souèges[489] and of an extensive physiological study by Rijven.[424] It is readily available for teaching purposes, and can easily be observed in the living state as well as in section. This can be done either by clearing whole ovules or even ovaries at different stages by warming in chloral hydrate, and observing the embryos through the cleared walls of the ovule, or by placing the ovules or ovaries in 5 per cent KOH for about 5 minutes, and then gently tapping the cover-slip. Intact embryos at various stages can be expelled from the embryo sac by this means and observed under the microscope (Fig. 8.4).

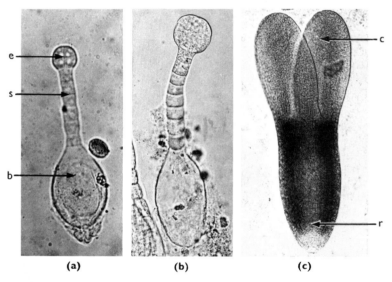

(a)   **(b)**   **(c)**

Fig. 8.4 Whole embryos of *Capsella bursa-pastoris*, expelled from the fertilized ovules. **(a)** Octant stage of the embryo (e). s, suspensor; b, basal cell.   × 100. **(b)** Late globular stage.   × 100. **(c)** Torpedo stage. c, cotyledon; r, radicle.   × 28.

The elegant ultrastructural work of Schultz and Jensen[455–458] has added greatly to our knowledge of embryo development in *Capsella*. In the embryo sac of *Capsella*, two **synergids** flank the egg cell at the micropylar end of the sac. The fine structure of the synergids suggests that they are metabolically much more active than the relatively quiescent egg cell. However, cytoplasmic connections exist between them. The synergids are believed to function in the absorption and transport of substances from the cells of the integuments to the egg cell. At the time of fertilization the pollen tube enters the embryo sac through the *filiform apparatus*, a modification of the wall of the synergids which results in a substantial increase in the surface area of the plasma membrane.[455] As in cotton, the egg cell is highly vacuolate. Histochemical tests show a pronounced positive reaction for protein and RNA in the cytoplasm of the egg cell; the nucleus stains intensely for protein and positively for DNA. Plastids are present round the egg nucleus, and the cytoplasm is packed with ribosomes. Few dictyosomes occur and the scant ER does not contain tubes as it does in cotton.[456] After fertilization the size of the vacuole decreases temporarily, but the size of the zygote does not decrease as it does in a pronounced way in cotton.[278] Polyribosomes appear in the cytoplasm, and the amount of lipid increases. The primary endosperm nucleus begins to divide before the zygote.[456]

Polarity becomes established in the egg cell early in development. The cytoplasm becomes aggregated at the chalazal end of the cell. In the zygote, this polarity is even more evident and a large vacuole develops at the micropylar end of the cell (see Part 1,[127] Fig. 2.1a). The first division of the zygote is transverse, i.e. at right angles to the axis of polarity of the cell, and unequal, resulting in the formation of a smaller *terminal* and a larger *basal cell*. Initially, histochemical staining indicates a higher concentration of protein and nucleic acids in the basal cell, but the situation is reversed after the next division of the basal cell.[456] Both cells have groups of lipid bodies and many small vacuoles (Fig. 8.5). The vacuoles in the basal cell enlarge until the cytoplasm is restricted to a peripheral area. At the micropylar end of the basal cell, ingrowths of the wall project into the cell lumen (Fig. 8.6a); these increase in size and number during embryo development. Dictyosome activity is associated with them. The plasma membrane follows the outline of the projections, resulting in greatly increased surface area.[456] Clearly, these projections strongly resemble those occurring in 'transfer cells', already discussed in Chapter 5 in relation to minor veins of the leaf. Their significance will be considered further below.

The basal cell divides transversely to give a 3-celled embryo, the 3rd cell being called the suspensor cell (Fig. 8.6a). The terminal cell also later divides transversely. The basal and suspensor cells eventually divide to give a *suspensor* of up to 10 cells, and the terminal cells divide to give quadrant, octant, 16- and 32-celled stages of the embryo. The protoderm of the embryo is established at the 16-celled stage. The globular embryo finally forms the primordia of two *cotyledons* (heart-shaped stage), and elongation of the embryo gives the torpedo stage. The embryo finally becomes a curved structure, which is sometimes called the walking-stick stage. Certain of these stages of development are illustrated in Figs. 8.1 and 8.4. Up to the globular stage of development, cleavage planes occur

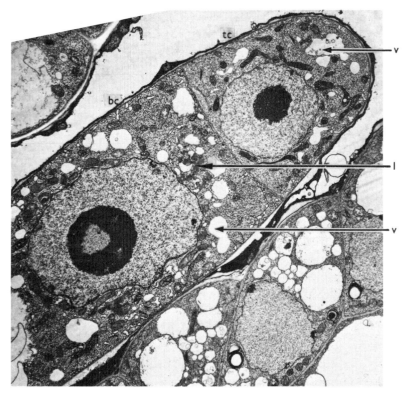

Fig. 8.5 Electron micrograph of a 2-celled embryo of *Capsella*. The terminal cell (tc) and the chalazal end of the basal cell (bc) are shown. Lipid bodies (l) and small vacuoles (v) are present in both cells. ×18,450. (From Schulz and Jensen,[456] Fig. 13, p. 815.)

with great regularity. Some of the older workers believed that divisions always occurred regularly in embryos, but later this approach was assigned less importance[553] and indeed in cotton the divisions are almost random.[278]

In the 16- and 32-celled stages of development of the embryo, higher concentrations of protein and nucleic acids become detectable, but there are not evident histochemical or ultrastructural differences between the component cells of the embryo, even though some of these now constitute the protoderm.[457] The density of ribosomes is greater in the cells of the embryo than in those of the suspensor, at the octant stage (Figs. 8.6b, 8.7, 8.8) and later. The distal cell of the suspensor, called the *hypophysis*, divides and the products of its daughter cell closest to the embryo later contribute to the root apex and root cap of the embryo. However, in their

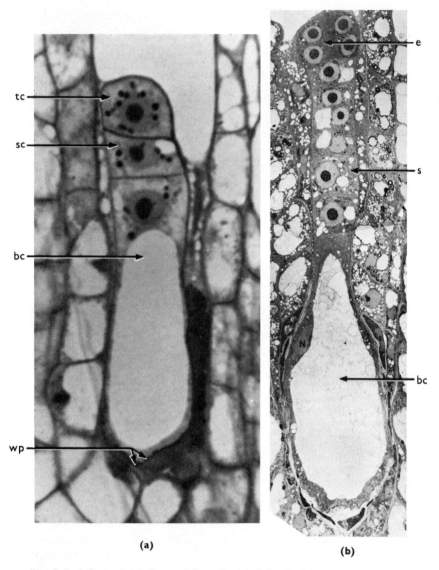

(a)   (b)

Fig. 8.6 L.S. young embryos of *Capsella*. (a) A 3-celled embryo, with terminal
cell (tc), suspensor cell (sc) and basal cell (bc). Darkly stained starch grains surround the nuclei. Wall projections (wp) are present at the micropylar end of the
basal cell.   ×1620. (From Schulz and Jensen,[456] Fig. 6, p. 809.) (b) Montage
of an octant embryo (e), with suspensor (s) and basal cell (bc). The nucleus (N)
of the basal cell is in the peripheral cytoplasm.   ×975. (From Schulz and Jensen,[458]
Fig. 1, p. 140.)

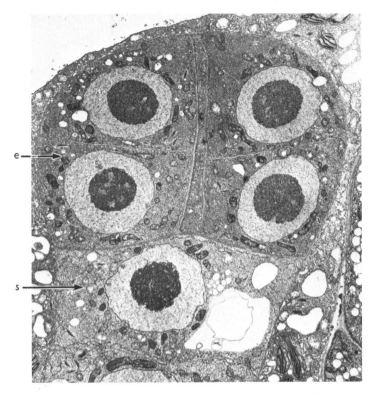

Fig. 8.7 Thin section of an octant embryo (e) of *Capsella* and the upper suspensor cells (s). The cells of the embryo are less vacuolate and have a greater density of ribosomes than the suspensor cells.   × 4810. (From Schulz and Jensen,[457] Fig. 5, p. 381.)

histochemistry and ultrastructure they resemble the cells of the suspensor (Fig. 8.9).

Two centres of growth become established in the globular embryo of dicotyledons, and localized cell division results in outgrowth of two cotyledon primordia (Fig. 8.1c). In this heart-shaped stage in *Capsella*, the cells are densely packed with ribosomes (Fig. 8.9). The whole embryo also stains intensely for protein and nucleic acids.[457]

Meanwhile various changes are also occurring in the cells of the suspensor. Plasmodesmata connect the embryo, suspensor and basal cells. By the time 10 cells are present in the suspensor, histochemical staining shows a lower concentration of protein and nucleic acids, and the density of ribosomes also decreases. By the heart-shaped stage of the embryo, degenera-

10

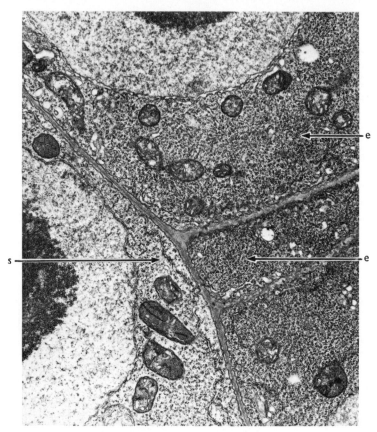

Fig. 8.8   Region where the octant embryo (e) of *Capsella* joins the suspensor (s), showing the differences in ribosome density between the cells of the two regions. × 14,700. (From Schulz and Jensen,[457] Fig. 7, p. 383.)

tion of the cytoplasm begins. Later, the suspensor becomes crushed by the growth of the embryo.[458]

At the heart-shaped stage the number of plasmodesmata in the end walls of the suspensor cells increases. Relatively few ribosomes remain in the cytoplasm. The lateral walls develop projections which protrude into the endosperm cells (Fig. 8.10).

The wall projections at the micropylar end of the basal cell also enlarge during development. There is an increase in protein in the cell.[458]

During later development the cotyledons and hypocotyl region of the embryo elongate, by means of transversely oriented divisions, and the

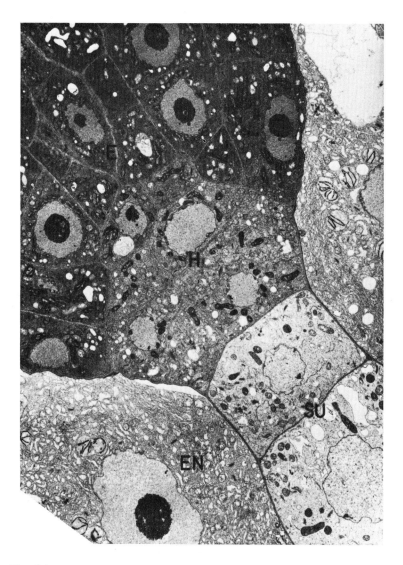

Fig. 8.9 Part of a heart-shaped embryo of *Capsella* showing the differences in structure and density between the embryo (e), suspensor (su) and hypophysis region (h). Well developed endosperm (en) is present.   ×2990. (From Schulz and Jensen,[457] Fig. 14, p. 389.)

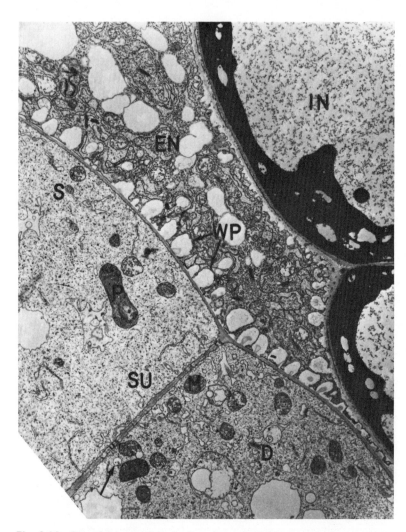

Fig. 8.10 Part of the suspensor (su), endosperm (en) and inner integument (in) of *Capsella* at the heart-shaped stage of embryo development. Projections (wp) are present on the outside of the lateral walls of the suspensor cells. Organelles such as dictyosomes (d), mitochondria (m), spherosomes (s), plastids (p) and multivesicular bodies (arrow) are still present in the cytoplasm of the suspensor cells but the density of ribosomes has decreased.   × 6860. (From Schulz and Jensen,[458] Fig. 8, p. 147.)

apices of root and shoot, and the intervening procambium become organized (Fig. 8.1d). As the embryo enlarges, it bends to accommodate itself to the shape of the embryo sac.

The suspensor cells retain a potentiality for growth. In experiments in which the developing embryos of *Eranthis hiemalis* were selectively killed, the surviving suspensor cells regenerated a new embryo. More than one adventive embryo could be formed in this way.[232]

Formerly it was believed that the suspensor served merely to push the developing embryo further into the nutritive endosperm tissue, although more recently an absorptive function was considered likely.[553] The ultrastructural evidence supports the view that both the suspensor and the basal cell function actively in absorption and transport of nutrients from the endosperm and integuments. The wall projections, which have numerous mitochondria associated with them, increase the absorptive surface.[458] In the embryo of *Stellaria*, also, it appears that the suspensor elaborates proteinaceous materials which are resorbed by the embryo during a period of rapid growth.[416] The wall outgrowths and ingrowths in *Capsella* bear a striking resemblance to those described for 'transfer cells' in various parts of the plant (see p. 169).[394] These cells are thought to be a device whereby the increase in surface to volume ratio of the protoplasts of parenchyma cells can increase their efficiency in the transport of solutes. On this view, the suspensor would be a specialized transport mechanism for the embryo, the wall projections increasing the efficiency of absorption from endosperm and other nutritive tissues, and the large number of plasmodesmata in the transverse walls contributing to free passage of materials along the suspensor to the embryo.

Experiments in which embryos of *Capsella* were supplied with ^{3}H-thymidine and ^{3}H-uridine showed that the nucleosides were not taken up by globular embryos, unless mannitol was added, suggesting that this addition led to a change in permeability. By the heart-shaped stage of development the embryo could incorporate the isotope.[407] Together with the evidence concerning the absorptive functions of the suspensor, this suggests that the failure of most embryos to grow in aseptic culture if removed from the embryo sac prior to the heart-shaped stage may be attributable not so much to a lack of necessary metabolites as to a lack of an efficient mechanism to absorb them. Further studies along these lines might be of great interest.

Embryogenesis in *Zea*

Embryogenesis in most monocotyledons is similar to that in dicotyledons up to the club-shaped or globular stage of development. With the inception of the cotyledon distinctive development ensues. The grass embryo is very complex, and possesses a number of structures the homology of which is in

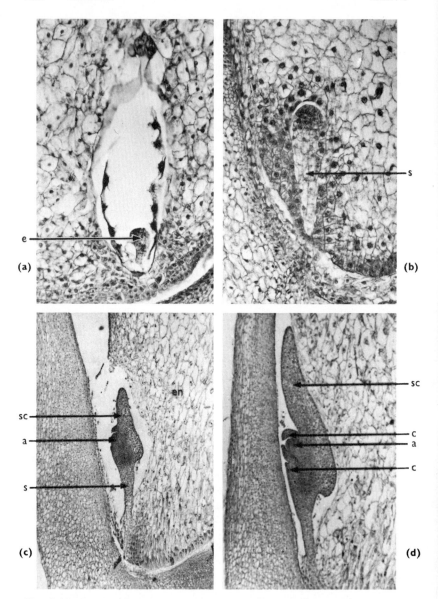

Fig. 8.11 Developing embryos of maize (corn), *Zea mays*. (a) and (b) Club-shaped embryos.   ×300. (c) and (d) The epicotyl apex (a) has been delimited. In (d) both the coleoptile (c) and first leaf are present.   ×45. e, embryo; en, endosperm; s, suspensor; sc, scutellum.

doubt (see Brown,[68] Wardlaw[553]). However, the importance of the cereals is such that the embryogenesis of *Zea* is considered briefly here.

A notch becomes apparent at one side of the more or less club-shaped embryo, and an outgrowth develops just above this; it constitutes the **coleoptile** primordium (Fig. 8.11). The tissue just below the notch becomes the shoot apex, and this is eventually encircled by the sheath of the coleoptile, here considered to be the first leaf. The upper part of the embryo develops rapidly into the **scutellum**, regarded as the cotyledon. The primary root may be considered to differentiate endogenously, the tissue external to it being the root sheath or **coleorhiza**.[68] The scutellum is appressed to the endosperm. The shoot apex gives rise to additional leaf primordia, and a short internode or **mesocotyl** develops between the points of attachment of the scutellum and coleoptile. Stages in the embryogenesis of maize (corn) are shown in Fig. 8.11 and of wheat in Fig. 8.12.

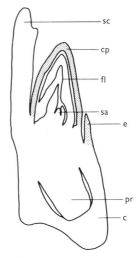

Fig. 8.12 Fully developed embryo of wheat, *Triticum aestivum*. c, coleorhiza; cp, coleoptile; e, epiblast; fl, foliage leaf; pr, primary root; sa, shoot apex; sc, scutellum. (From Brown,[68] Fig. 1, p. 219.)

The epiblast (Fig. 8.12) has been variously interpreted, often as a cotyledon or part of one.[68]

The single cotyledon in monocotyledons sometimes appears terminal, and the apical meristem lateral. Evidence from dicotyledons which develop only one cotyledon supports the view that the cotyledon is a lateral structure, however. It has recently been suggested that several species of the monocotyledon *Dioscorea* have two cotyledons, one of which has an absorbing function and remains within the endosperm, while the other emerges and functions as a leaf.[327]

The coleoptile has been extensively utilized in auxin studies, and is still used in one of the standard tests for auxin activity.[330]

Various workers have successfully transplanted embryos of one cereal into the endosperm of another species. In some such transplants better growth of the embryo can be obtained. Indeed, it appears that the endosperm can have interesting effects on the subsequent metabolism of the embryos. This work is discussed further by Wardlaw.[561]

EMBRYOIDS

The formation of embryo-like structures, or **embryoids**, from individual or small groups of cultured cells of carrot root phloem, carrot embryos or

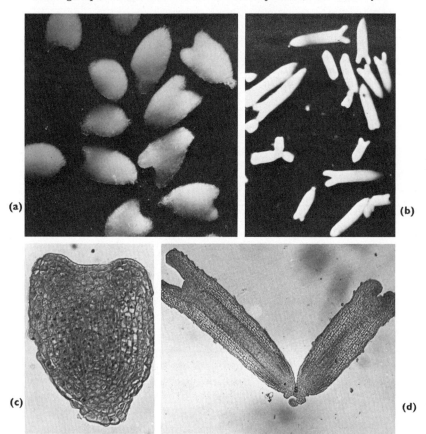

Fig. 8.13 Embryoids of wild carrot. (a) Heart stage of development. × 60. (b) Older embryos. × 27. (c) Early heart stage in L.S. × 280. (d) Two older embryos joined by a common 'suspensor' (s). × 55. (From Halperin,[236] Figs. 8, 9, 21 and 35, pp. 445, 448 and 449.)

pieces of petiole has already been considered to some degree in Chapter 2 of Part 1,[127] and is more fully discussed in the volume by Street and Öpik[497] in this series. Some of these structures most strikingly resemble normal embryos (Fig. 8.13). Haccius[233] has raised the question of whether embryoids have a suspensor, and has concluded that since an individual suspensor cannot be delimited they should properly be regarded as adventive embryos, without full identity with zygotic embryos. However, Halperin[236] has demonstrated suspensor-like regions in carrot embryoids

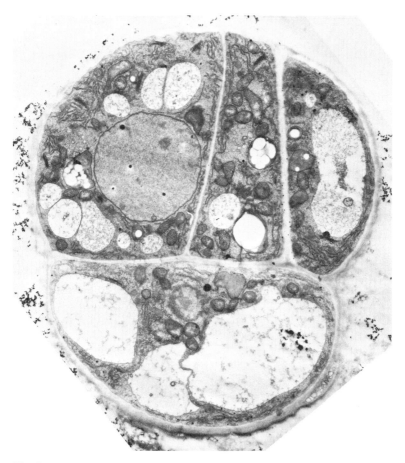

Fig. 8.14 Four-celled structure resembling a proembryo from a sieved suspension of cells of wild carrot. There is a large, vacuolate, suspensor-like cell and 3 outer, more meristematic cells. × 6820. (From Halperin and Jensen,[237] Fig. 19, p. 440. Copyright Academic Press.)

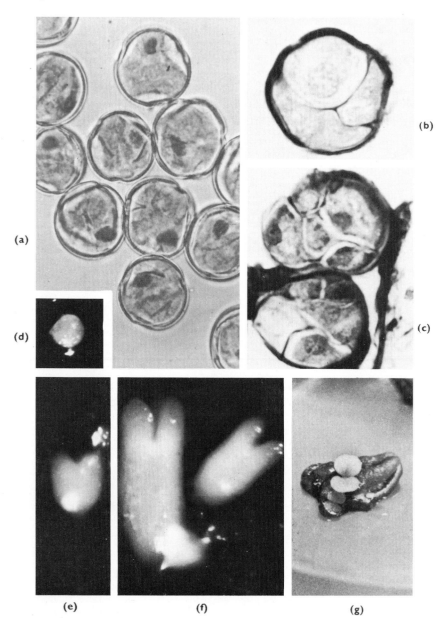

Fig. 8.15 Embryoids in tobacco. (a) Pollen grains at the time of culturing, stained by the Feulgen technique.   ×1500. (b) Pollen grain in a cultured anther

(Fig. 8.13d). These areas sometimes store starch and are also the region where multiple embryoids are united. Although they may perhaps fulfil the function of suspensors, they are certainly not morphologically identical to the suspensors of zygotic embryos. However, electron microscopic studies of carrot embryoids reveal suspensor-like regions, more highly vacuolated than the proembryo-like structure at the other end of the embryoid, in quite young embryoids (Fig. 8.14).[237] Thus polarization seems to occur in these structures at an early stage of development.

Embryo-like developments of this type have now been obtained from many different species and somatic tissues (see Wardlaw).[559, 561] Embryoids from the flower stalks of *Ranunculus sceleratus* even develop from epidermal cells that are still attached to the plant.[316] More recently, embryoids have been obtained from the pollen of *Datura*[222, 223] and tobacco.[383] In tobacco many embryoids could be readily obtained, if anthers from flowers at an appropriate stage of development were cultured on a medium containing sucrose. The pollen grains had to be uninucleate and devoid of starch. The embryoids passed through globular, heart-shaped and torpedo stages of development (Fig. 8.15) and eventually formed normal plants with flowers which, however, did not set seed. These embryoids from pollen are especially interesting because they develop from haploid cells, are themselves haploid, and give rise to haploid plants. This technique is consequently of great potential value to plant breeders.

These studies of embryoids are of considerable significance because they indicate that a single cell from whatever source, whether haploid or diploid, often follows a course of development similar to that of the angiosperm zygote. Not all isolated cells develop in this way, however, notably those developing from mesophyll, discussed in Chapter 5.[294, 295] It is not difficult to envisage the formation of a spherical ball of cells from a single cell, but the establishment of polarity, leading to the differentiation of root and shoot apices, and the formation of two 'cotyledons' with such regularity is more difficult to understand in cells growing freely in a culture medium. Perhaps further experimental study of these induced embryoids will contribute to a greater knowledge of the factors involved in normal embryogenesis.

---

showing 3 cells inside the exine.   × 920. (c) Young embryos in a cultured anther. × 920. (d) A globular stage.   × 135. (e) Heart-shaped stage of embryoid development.   × 135. (f) Torpedo stage.   × 135. (g) Plantlets emerging from a cultured anther. Note the two cotyledons.   × 5. (By courtesy of Dr. J. P. Nitsch. (a)–(c) from Nitsch,[382a] Figs. 5–7, p. 396; (e) and (f) from Nitsch and Nitsch,[383] Figs. 3D and E, p. 86, Copyright 1969 by the American Association for the Advancement of Science.)

Appendix

CLASS EXPERIMENTS

The following experiments have been adapted for class use.* Detailed instructions follow.

The control of cambial activity in woody stems

The aim of this experiment is to simulate the effects of actively growing buds on the activity of the vascular cambium and the differentiation of its products by supplying various growth hormones separately and in combination. The hormones are applied in a lanolin paste to disbudded woody twigs.

To make the lanolin paste, weigh out a quantity of anhydrous lanolin (in the example given, about 60 g) in a beaker of known weight. Melt the lanolin overnight in an oven at about 60°C. To the deep yellow lanolin, add water to make 100 g, stirring vigorously. The lanolin will turn almost white. Allow it to stand overnight at room temperature; next day, shake out any liquid which is not incorporated into the lanolin mixture. Dissolve the appropriate amount of crystalline hormone in a small amount of absolute alcohol and add this to the warmed hydrous lanolin. Stir very vigorously to ensure even distribution of the hormone. To make the control lanolin paste, add a similar quantity of absolute alcohol, without hormone, to the hydrous lanolin.

Woody twigs of *Populus*, *Acer* or *Fraxinus* are suitable materials; results with *Populus* spp. have been especially good. These twigs should be in a non-dormant condition, i.e. not completely dormant, but obtained before the buds have burst. The twigs should be vigorous one-year-old shoots which have grown about 30–50 cm in the previous year, and should be about 0·7–1·0 cm in diameter. If necessary, they can be stored in a cold room (*c.* 5°C) for a period prior to use.

* I am indebted to Professor P. F. Wareing for some further details on the experiment on the woody stems, which is based on published work by him and his co-workers.[566, 568] Some additional modifications are also suggested.

The twigs should be cut into 15 cm lengths (avoiding the thinner parts at the tips of the twigs) and all lateral buds removed. During the course of the experiment it may be necessary to remove additional small buds which have been overlooked or improperly excised. Stand the twigs in jars of water and apply the lanolin preparations to the upper (apical) cut end of each twig. Since the occasional formation of roots near the base of hormone-treated twigs during the period of the experiments suggests that the substances may move down the twig and out into the water, it is important that twigs given different treatments should be placed in separate jars.

Treat four twigs of the same species as follows:

(1) Plain lanolin (control)
(2) 500 ppm indoleacetic acid (IAA) in lanolin
(3) 500 ppm gibberellic acid (GA) in lanolin
(4) IAA/GA mixture in lanolin, both at 500 ppm

Apply the lanolin generously to the freshly cut apical end of each twig, ensuring that the cut surface is fully covered. Replace every 3–4 days, removing a *thin* slice of twig below the old application before applying the fresh lanolin. The old lanolin can be removed with Kleenex or Kimwipes. Intervals of 2 and 5 days between applications are also satisfactory, but applications should be renewed twice weekly on some convenient time schedule. Lanolin preparations should be kept in a refrigerator, and removed an hour or two before required. Individual twigs can be conveniently labelled with price tags or similar small labels.

A convenient method for removing slices from the ends of the twigs is to place the twig in a horizontal position on a paper towel on the bench surface with several inches of the basal end projecting over the edge of the bench. Place a sharp single-edged razor blade near the apical end in the desired position and hold it with the right hand (if right-handed); meanwhile hold the piece of twig with the left hand and rotate it. This method is especially useful whenever longer pieces of twig are to be excised, e.g. at the beginning of the experiment.

After 3 weeks, transverse sections should be cut at a distance of about 1 cm or slightly less below the treated end of each twig. Sections can be cut fairly easily with *sharp* razors or razor blades, or on a sliding microtome. Excellent differential staining can be achieved by placing the sections for a few minutes in a solution of toluidine blue,[390] washing in water and then mounting in water or dilute glycerine. This stains the lignified elements of the secondary xylem a greenish-blue, and the phloem, cambium and undifferentiated cambial region purple.

The control of vascular cambial activity in herbaceous stems

Plants of *Phaseolus vulgaris* approximately 14 days old (depending on the growing conditions) are suitable material. Decapitate a plant above a suitable internode (probably the top one which has extended). Cut transverse sections of this representative internode and stain in toluidine blue.[390] Examine for the presence of vascular cambium and secondary vascular tissues. The internodes to be treated should have no differentiated secondary tissues. In 14-day-old plants the internode below the first trifoliate leaf (see Fig. 4.8a) is usually suitable.

Having selected the internodes to be treated, decapitate the plants above this internode. Keep some plants intact to serve as controls. Apply colour-coded Beem capsules* containing the following solutions to the decapitated plants:

(1) Distilled water in 1 per cent agar (control)
(2) 10 mg/l indoleacetic acid (IAA) in 1 per cent agar
(3) 10 mg/l gibberellic acid (GA) in 1 per cent agar
(4) 10 mg/l IAA + 10 mg/l GA in 1 per cent agar

After 3 or 4 days, remove the capsules (the agar will have dried up considerably), make a fresh cut across the surface of the stem and apply a fresh capsule of the same kind.

After 1 week, make transverse sections of each treated internode a short distance below the cut surface and stain in toluidine blue. Compare the effects of the various treatments on cambial activity and the differentiation of its products.

A possible additional treatment is the application of 2,3,5-tri-iodobenzoic acid (TIBA) in lanolin around the apical end of the same internode in intact plants. Since this substance inhibits the polar transport of auxin from the apical parts, a result comparable to that in decapitated plants treated with GA is obtained. Concentrations of about 0·75 to 1·0 per cent TIBA in lanolin appear to be suitable. However, these concentrations also inhibit the growth of the apical parts of the shoot, which is an undesirable effect making the results harder to interpret. Some plants give a better reaction than others. However, this addition to the experiment seems to be worth following up further.

* These are plastic capsules used for embedding material for electron microscopy. They can be conveniently colour-coded by applying coloured dots such as are used to mark projection slides to the hinged lid of the capsule, which is left open after application.

Further Reading

CARLQUIST, S. (1961). *Comparative Plant Anatomy*. Holt, Rinehart and Winston, New York.

CLOWES, F. A. L. (1961). *Apical Meristems*. Blackwell, Oxford.

CLOWES, F. A. L. and JUNIPER, B. E. (1968). *Plant Cells*. Blackwell, Oxford.

CUTLER, D. F. (1969). *Anatomy of the Monocotyledons*. IV. *Juncales*. General editor, C. R. Metcalfe. Clarendon Press, Oxford.

CUTTER, E. G. (1969). *Plant Anatomy: Experiment and Interpretation*. Part 1. *Cells and Tissues*. Edward Arnold, London.

DAVIS, G. L. (1966). *Systematic Embryology of the Angiosperms*. Wiley, New York.

EAMES, A. J. (1961). *Morphology of the Angiosperms*. McGraw-Hill, New York.

EAMES, A. J. and MACDANIELS, L. H. (1947). *An Introduction to Plant Anatomy*. 2nd edition. McGraw-Hill, New York and London.

ESAU, K. (1960). *Anatomy of Seed Plants*. Wiley, New York.

ESAU, K. (1965). *Plant Anatomy*. 2nd edition. Wiley, New York.

ESAU, K. (1965). *Vascular Differentiation in Plants*. Holt, Rinehart and Winston, New York.

FAHN, A. (1967). *Plant Anatomy*. (Translated from the Hebrew by Sybil Broido-Altman.) Pergamon Press, Oxford.

FOSTER, A. S. (1949). *Practical Plant Anatomy*. 2nd edition. Van Nostrand, New York.

MAHESHWARI, P. (1950). *An Introduction to the Embryology of Angiosperms*. McGraw-Hill, New York.

METCALFE, C. R. (1960). *Anatomy of the Monocotyledons*. I. *Gramineae*. Clarendon Press, Oxford.

METCALFE, C. R. and CHALK, L. (1950). *Anatomy of the Dicotyledons*. Vols. I and II. Clarendon Press, Oxford.

TOMLINSON, P. B. (1961). *Anatomy of the Monocotyledons*. II. *Palmae*. General editor, C. R. Metcalfe. Clarendon Press, Oxford.

TOMLINSON, P. B. (1969). *Anatomy of the Monocotyledons*. III. *Commelinales–Zingiberales*. General editor, C. R. Metcalfe. Clarendon Press, Oxford.

WARDLAW, C. W. (1955). *Embryogenesis in Plants*. Methuen, London.

WARDLAW, C. W. (1965). *Organization and Evolution in Plants*. Longmans, London.

WARDLAW, C. W. (1968). *Morphogenesis in Plants*. Methuen, London.

References

1. ABBE, E. C., PHINNEY, B. O. and BAER, D. F. (1951). The growth of the shoot apex in maize: internal features. *Am. J. Bot.*, **38**, 744–751.
2. ADDICOTT, F. T. (1965). Physiology of abscission. *Handb. PflPhysiol.*, **15, 2**, 1094–1126.
3. AGHION, D. (1962). Conditions expérimentales condiusant à l'initiation et au dévelloppement de fleurs à partir de la culture stérile de fragments de tige de tabac. *C. r. hebd. Séanc. Acad. Sci., Paris*, **255**, 993–995.
4. ALFIERI, I. R. and EVERT, R. F. (1968). Analysis of meristematic activity in the root tip of *Melilotus alba* Desr. *New Phytol.*, **67**, 641–647.
5. ALLSOPP, A. (1953). Experimental and analytical studies of pteridophytes. XXI. Investigations on *Marsilea*. 3. The effect of various sugars on development and morphology. *Ann. Bot.*, N.S., **17**, 447–463.
6. ALLSOPP, A. (1962). The effects of gibberellic acid on morphogenesis in *Marsilea drummondii* A. Br. *Phytomorphology*, **12**, 1–10.
7. ALLSOPP, A. (1963). Morphogenesis in *Marsilea*. *J. Linn. Soc. (Bot.)*, **58**, 417–427.
8. ALLSOPP, A. (1964). Shoot morphogenesis. *A. Rev. Pl. Physiol.*, **15**, 225–254.
9. ALLSOPP, A. (1965a). The significance for development of water supply, osmotic relations and nutrition. *Handb. PflPhysiol.*, **15, 1**, 504–552.
10. ALLSOPP, A. (1965b). Heteroblastic development in cormophytes. *Handb. PflPhysiol.*, **15, 1**, 1172–1221.
11. ALLSOPP, A. (1965c). Land and water forms: physiological aspects. *Handb. PflPhysiol.*, **15, 1**, 1236–1255.
12. ALLSOPP, A. (1967). Heteroblastic development in vascular plants. *Adv. Morphogen.*, **6**, 127–171.
13. AMIR, J. (1969). A study on the reproductive stage of the groundnut *Arachis hypogaea* L. Induction of pod setting in the upper-nodal gynophores. *Ann. Bot.*, N.S., **33**, 333–338.
14. ANGOLD, R. E. (1968). The formation of the generative cell in the pollen grain of *Endymion non-scriptus* (L.). *J. Cell Biol.*, **3**, 573–578.
15. ANKER, L. (1968). On gravi-sensitivity in plants. *Acta bot. neerl.*, **17**, 385–389.
16. ARBER, A. (1946). Goethe's botany. *Chron. Bot.*, **10**, 63–126.
17. ARNOLD, C. A. (1940). A note on the origin of the lateral rootlets of *Eichhornia crassipes* (Mart.) Solms. *Am. J. Bot.*, **27**, 728–730.

18. ARNOTT, H. J. (1962). The seed, germination, and seedling of *Yucca*. *Univ. Calif. Publs Bot.*, **35**, 1–164.

19. ARZEE, T., LIPHSCHITZ, N. and WAISEL, Y. (1968). The origin and development of the phellogen in *Robinia pseudacacia* L. *New Phytol.*, **67**, 87–93.

20. AVERY, G. S., Jr. (1933). Structure and development of the tobacco leaf. *Am. J. Bot.*, **20**, 565–592.

21. BACQUAR, S. R. and AFAC HUSAIN, S. (1969). Cytoplasmic channels and chromatin migration in the meiocytes of *Arnebia hispidissima* (Sieb.) DC. *Ann. Bot.*, *N.S.*, **33**, 821–831.

22. BALATINECZ, J. J. and FARRAR, J. L. (1966). Pattern of renewed cambial activity in relation to exogenous auxin in detached woody shoots. *Can. J. Bot.*, **44**, 1108–1110.

23. BALATKOVÁ, V. and TUPÝ, J. (1968). Test-tube fertilization in *Nicotiana tabacum* by means of an artificial pollen tube culture. *Biologia Pl.*, **10**, 266–270.

24. BALFOUR, E. (1958). The development of the vascular systems of *Macropiper excelsum* Forst. II. The mature stem. *Phytomorphology*, **8**, 224–233.

25. BALL, E. (1941). The development of the shoot apex and of the primary thickening meristem in *Phoenix canariensis* Chanb., with comparisons to *Washingtonia filifera* Wats. and *Trachycarpus excelsa* Wendl. *Am. J. Bot.*, **28**, 820–832.

26. BALL, E. (1952). Morphogenesis of shoots after isolation of the shoot apex of *Lupinus albus*. *Am J. Bot.*, **39**, 167–191.

27. BALL, E. (1955). On certain gradients in the shoot tip of *Lupinus albus*. *Am. J. Bot.*, **42**, 509–521.

28. BALL, E. (1956). Growth of the embryo of *Ginkgo biloba* under experimental conditions. II. Effects of a longitudinal split in the tip of the hypocotyl. *Am. J. Bot.*, **43**, 802–810.

29. BALL, E. (1960). Cell divisions in living shoot apices. *Phytomorphology*, **10**, 377–396.

30. BARBER, J. T. and STEWARD, F. C. (1968). The proteins of *Tulipa* and their relation to morphogenesis. *Devl Biol.*, **17**, 326–349.

31. BARKER, W. G. (1969). Growth and development of the banana plant. Gross leaf emergence. *Ann. Bot.*, *N.S.*, **33**, 523–535.

32. BARLOW, P. W. (1969). Cell growth in the absence of division in a root meristem. *Planta*, **88**, 215–223.

33. BARNELL, E. (1939). Studies in tropical fruits. V. Some anatomical aspects of fruit-fall in two tropical arboreal plants. *Ann. Bot.*, *N.S.*, **3**, 77–89.

34. BARNELL, H. R. and BARNELL, E. (1945). Studies in tropical fruits. XVI. The distribution of tannins within the banana and the changes in their condition and amount during ripening. *Ann. Bot.*, *N.S.*, **9**, 77–99.

35. BASFORD, K. H. (1961). Morphogenetic responses to gibberellic acid of a radiation-induced mutant dwarf in groundsel, *Senecio vulgaris* L. *Ann. Bot.*, *N.S.*, **25**, 279–302.

36. BAYER, D. E., FOY, C. L., MALLORY, T. E. and CUTTER, E. G. (1967). Morphological and histological effects of trifluralin on root development. *Am. J. Bot.*, **54**, 945–952.

37. BECKER, D. A. (1968). Stem abscission in the tumbleweed, *Psoralea*. *Am. J. Bot.*, **55**, 753–756.
38. BELL, P. R. and WOODCOCK, C. L. F. (1968). *The Diversity of Green Plants.* Edward Arnold, London.
38a. BERNIER, G. (1964). Étude histophysiologique et histochimique de l'évolution du méristème apicale de *Sinapis alba* L., cultivé en milieu conditionné et en diverses durées de jour favorables ou défavorables à la mise à fleurs. *Mem. Acad. r. Belg. Cl. Sci.* 4°, **16**, 1–149.
39. BERNIER, G. (1966). The morphogenetic role of the apical meristem in higher plants. *Les Congrés et Colloques de l'Université de Liège*, **38**, Les Phytohormones et l'Organogenèse, 151–211.
40. BERNIER, G. (1969). *Sinapis alba* L. In *The Induction of Flowering*, EVANS, L. T., 305–327. Cornell Univ. Press, Ithaca, New York.
41. BERNIER, G., BRONCHART, R. and JACQMARD, A. (1964). Action of gibberellic acid on the mitotic activity of the different zones of the shoot apex of *Rudbeckia bicolor* and *Perilla nankinensis*. *Planta*, **61**, 236–244.
42. BHAR, D. S. and RADFORTH, N. W. (1969). Vegetative and reproductive development of shoot apices of *Pharbitis nil* as influenced by photoperiodism. *Can. J. Bot.*, **47**, 1403–1406.
43. BIENIEK, M. E. and MILLINGTON, W. F. (1967). Differentiation of lateral shoots as thorns in *Ulex europaeus*. *Am J. Bot.*, **54**, 61–70.
44. BIENIEK, M. E. and MILLINGTON, W. F. (1968). Thorn formation in *Ulex europaeus* in relation to environmental and endogenous factors. *Bot. Gaz.*, **129**, 145–150.
45. BISALPUTRA, T., DOWNTON, W. J. S. and TREGUNNA, E. B. (1969). The distribution and ultrastructure of chloroplasts in leaves differing in photosynthetic carbon metabolism. I. Wheat, *Sorghum* and *Aristida* (Gramineae). *Can. J. Bot.*, **47**, 15–21.
46. BLOCH, R. (1946). Differentiation and pattern in *Monstera deliciosa*. The idioblastic development of the trichosclereids in the air root. *Am. J. Bot.*, **33**, 544–551.
47. BOKE, N. H. (1940). Histogenesis and morphology of the phyllode in certain species of *Acacia*. *Am. J. Bot.*, **27**, 73–90.
48. BOKE, N. H. (1948). Development of the perianth in *Vinca rosea* L. *Am. J. Bot.*, **35**, 413–423.
49. BOKE, N. H. (1949). Development of the stamens and carpels in *Vinca rosea* L. *Am. J. Bot.*, **36**, 535–547.
50. BONNETT, H. T., Jr. (1968). The root endodermis: fine structure and function. *J. Cell Biol.*, **37**, 199–205.
51. BONNETT, H. T., Jr. (1969). Cortical cell death during lateral root formation. *J. Cell Biol.*, **40**, 144–159.
52. BONNETT, H. T., Jr. and TORREY, J. G. (1966). Comparative anatomy of endogenous bud and lateral root formation in *Convolvulus arvensis* roots cultured *in vitro*. *Am. J. Bot.*, **53**, 496–507.
53. BOOTH, A., MOORBY, J., DAVIES, C. R., JONES, H. and WAREING, P. F. (1962). Effects of indolyl-3-acetic acid on the movement of nutrients within plants. *Nature, Lond.*, **194**, 204–205.
54. BORCHERT, R. (1965). Gibberellic acid and rejuvenation of apical meristems in *Acacia melanoxylon*. *Naturwissenschaften*, **52**, 65–66.
55. BORNMAN, C. H. (1967a). The relationship between tylosis and abscission in cotton (*Gossypium hirsutum* L.) explants. *S. Afr. J. agric. Sci.*, **10**, 143–154.

56. BORNMAN, C. H. (1967b). Some ultrastructural aspects of abscission in *Coleus* and *Gossypium*. *S. Afr. J. Sci.*, **63**, 325–331.

57. BORNMAN, C. H., ADDICOTT, F. T., LYON, J. L., and SMITH, O. E. (1968). Anatomy of gibberellin-induced stem abscission in cotton. *Am. J. Bot.*, **55**, 369–375.

58. BORNMAN, C. H., SPURR, A. R. and ADDICOTT, F. T. (1967). Abscisin, auxin, and gibberellin effects on the developmental aspects of abscission in cotton (*Gossypium hirsutum*). *Am. J. Bot.*, **54**, 125–135.

59. BORNMAN, C. H., SPURR, A. R. and ADDICOTT, F. T. (1969). Histochemical localization by electron microscopy of pectic substances in abscising tissue. *Jl S. Afr. Bot.*, **35**, 253–264.

60. BOSTRACK, J. M. and MILLINGTON, W. F. (1962). On the determination of leaf form in an aquatic heterophyllous species of *Ranunculus*. *Bull. Torrey bot. Club*, **89**, 1–20.

61. BOWES, B. G. (1961). Inequality in the development of the axillary members in *Glechoma hederacea* L. *Ann. Bot.*, *N.S.*, **25**, 391–406.

62. BOWES, B. G. (1963). The structure and development of the vegetative shoot apex in *Glechoma hederacea* L. *Ann. Bot.*, *N.S.*, **27**, 357–364.

63. BOWES, B. G. (1965). The ultrastructure of the shoot apex and young shoot of *Glechoma hederacea* L. *Cellule*, **65**, 351–356.

64. BRINK, R. A. and COOPER, D. C. (1947). The endosperm in seed development. *Bot. Rev.*, **13**, 423–541.

65. BROWN, C. L. (1964). The influence of external pressure on the differentiation of cells and tissues cultured *in vitro*. In *The Formation of Wood in Forest Trees*, ZIMMERMANN, M. H., 389–404. Academic Press, New York.

66. BROWN, C. L. and SAX, K. (1962). The influence of pressure on the differentiation of secondary tissues. *Am. J. Bot.*, **49**, 683–691.

67. BROWN, J. A. M., MIKSCHE, J. P. and SMITH, H. H. (1964). An analysis of H^3-thymidine distribution throughout the vegetative meristem of *Arabidopsis thaliana* (L.) Heynh. *Rad. Bot.*, **4**, 107–113.

68. BROWN, W. V. (1960). The morphology of the grass embryo. *Phytomorphology*, **10**, 215–223.

69. BRUMFIELD, R. T. (1943). Cell-lineage studies in root meristems by means of chromosome rearrangements induced by X-rays. *Am. J. Bot.*, **30**, 101–110.

70. BÜNNING, E. (1952a). Morphogenesis in plants. *Surv. biol. Prog.*, **2**, 105–140.

71. BÜNNING, E. (1952b). Weitere Untersuchungen über die Differenzierungsvorgänge in Wurzeln. *Z. Bot.*, **40**, 385–406.

72. BUTCHER, D. N. and STREET, H. E. (1964). Excised root culture. *Bot. Rev.*, **30**, 513–586.

73. BUTLER, R. D. and LANE, G. R. (1959). The study of apical development in relation to etiolation. *J. Linn. Soc. (Bot.)*, **56**, 170–176.

74. BUVAT, R. (1952). Structure, évolution et fonctionnement du méristème apical de quelques dicotylédones. *Annls Sci. nat. (Bot.)*, Sér. 11, **13**, 199–300.

75. BUVAT, R. (1955). Le méristème apical de la tige. *Année biol.*, **31**, 596–656.

76. BYSTROM, B. G., GLATER, R. B., SCOTT, F. M. and BOWLER, E. S. C. (1968). Leaf surface of *Beta vulgaris*—electron microscope study. *Bot. Gaz.*, **129**, 133–138.

77. CARR, D. J. and PATE, J. S. (1967). Ageing in the whole plant. *Symp. Soc. exp. Biol.*, **21**, 559–599.

78. CARUSO, J. L. and CUTTER, E. G. (1966). Proliferation of cells in the central cylinder of the reduced mutant in lanceolate tomato. *Science, N.Y.*, **154**, 1021–1023.

79. CARUSO, J. L. and CUTTER, E. G. (1970). Morphogenetic aspects of a leafless mutant in tomato. II. Induction of a vascular cambium. *Am. J. Bot.*, **57**, 420–429.

80. CASPERSON, G. (1965). Über endogene Faktoren der Reaktionsholzbildung. I. Wuchsstoffapplikation an Kastanienepikotylen. *Planta*, **64**, 225–240.

81. CASPERSON, G. (1968). Wirkung von Wuchs- und Hemmstoffen auf die Kambiumtätigkeit und Reaktionsholzbildung. *Physiologia Pl.*, **21**, 1312–1321.

82. CHEADLE, V. I. (1937). Secondary growth by means of a thickening ring in certain monocotyledons. *Bot. Gaz.*, **98**, 535–555.

83. CHOUARD, P. (1936). La nature et le rôle des formations dites 'secondaires' dans l'édification de la tige des monocotylédones. *Bull. bot. Soc. Fr.*, **83**, 819–836.

84. CHURCH, A. H. (1904). *On the Relation of Phyllotaxis to Mechanical Laws*. Williams and Norgate, London.

84a. CHURCH, A. H. (1920). *On Problems of Phyllotaxis*. Oxford bot. Mem. no. 6.

85. CLOWES, F. A. L. (1950). Root apical meristems of *Fagus sylvatica*. *New Phytol.*, **49**, 248–268.

86. CLOWES, F. A. L. (1953). The cytogenerative centre in roots with broad columellas. *New Phytol.*, **52**, 48–57.

87. CLOWES, F. A. L. (1954). The promeristem and the minimal constructional centre in grass root apices. *New Phytol.*, **53**, 108–116.

88. CLOWES, F. A. L. (1956a). Nucleic acids in root apical meristems of *Zea*. *New Phytol.*, **55**, 29–34.

89. CLOWES, F. A. L. (1956b). Localization of nucleic acid synthesis in root meristems. *J. exp. Bot.*, **7**, 307–312.

90. CLOWES, F. A. L. (1957). Chimeras and meristems. *Heredity*, **11**, 141–148.

91. CLOWES, F. A. L. (1958a). Development of quiescent centres in root meristems. *New Phytol.*, **57**, 85–88.

92. CLOWES, F. A. L. (1958b). Protein synthesis in root meristems. *J. exp. Bot.*, **9**, 229–238.

93. CLOWES, F. A. L. (1959a). Apical meristems of roots. *Biol. Rev.*, **34**, 501–529.

94. CLOWES, F. A. L. (1959b). Reorganization of root apices after irradiation. *Ann. Bot., N.S.*, **23**, 205–210.

95. CLOWES, F. A. L. (1959c). Adenine incorporation and cell division in shoot apices. *New Phytol.*, **58**, 16–19.

96. CLOWES, F. A. L. (1961a). *Apical Meristems*. Blackwell, Oxford.

97. CLOWES, F. A. L. (1961b). Duration of the mitotic cycle in a meristem. *J. exp. Bot.*, **12**, 283–293.

98. CLOWES, F. A. L. (1961c). Effects of β-radiation on meristems. *Expl Cell Res.*, **25**, 529–534.

99. CLOWES, F. A. L. (1962). Rates of mitosis in a partially synchronous meristem. *New Phytol.*, **61**, 111–118.

100. CLOWES, F. A. L. (1964). The quiescent centre in meristems and its behaviour after irradiation. *Brookhaven Symp. Biol.*, **16**, 46–58.

101. CLOWES, F. A. L. (1965a). Synchronization in a meristem by 5-aminouracil. *J. exp. Bot.*, **16**, 581–586.

102. CLOWES, F. A. L. (1965b). Meristems and the effect of radiation on cells. *Endeavour*, **24**, 8–12.

103. CLOWES, F. A. L. (1965c). The duration of the G_1 phase of the mitotic cycle and its relation to radiosensitivity. *New Phytol.*, **64**, 355–359.

104. CLOWES, F. A. L. (1968). The DNA content of the cells of the quiescent centre and root cap of *Zea mays. New Phytol.*, **67**, 631–639.

105. CLOWES, F. A. L. and JUNIPER, B. E. (1964). The fine structure of the quiescent centre and neighbouring tissues in root meristems. *J. exp. Bot.*, **15**, 622–630.

106. CLOWES, F. A. L. and STEWART, H. E. (1967). Recovery from dormancy in roots. *New Phytol.*, **66**, 115–123.

107. COGGINS, C. W., Jr. and HIELD, H. Z. (1968). Plant-growth regulators. In *The Citrus Industry*, II, REUTHER, W., BATCHELOR, L. D. and WEBBER, H. J., 371–389. Univ. California, Div. Agric. Sci.

108. COOK, C. D. K. (1969). On the determination of leaf form in *Ranunculus aquatilis. New Phytol.*, **68**, 469–480.

109. COOKE, G. B. (1948). Cork and cork products. *Econ. Bot.*, **2**, 393–402.

110. CORNER, E. J. H. (1951). The leguminous seed. *Phytomorphology*, **1**, 117–150.

111. CORSON, G. E., Jr. (1969). Cell division studies of the shoot apex of *Datura stramonium* during transition to flowering. *Am. J. Bot.*, **56**, 1127–1134.

112. CORSON, G. E., Jr. and GIFFORD, E. M., Jr. (1969). Histochemical studies of the shoot apex of *Datura stramonium* during transition to flowering. *Phytomorphology*, **19**, 189–196.

113. CÔTÉ, W. A., Jr. and DAY, A. C. (1965). Anatomy and ultrastructure of reaction wood. In *Cellular Ultrastructure of Woody Plants*, CÔTÉ, W. A., Jr., 391–418. Syracuse University Press.

114. CRANE, J. C. and PUNSRI, P. (1956). Comparative growth of the endosperm and the embryo in unsprayed and 2,4,5-trichlorophenoxyacetic acid sprayed Royal and Tilton apricots. *Proc. Am. Soc. hort. Sci.*, **68**, 96–104.

115. CRONSHAW, J. and MOREY, P. R. (1965). Induction of tension wood by 2,3,5-tri-iodobenzoic acid. *Nature, Lond.*, **205**, 816–818.

116. CRONSHAW, J. and MOREY, P. R. (1968). The effect of plant growth substances on the development of tension wood in horizontally inclined stems of *Acer rubrum* seedlings. *Protoplasma*, **65**, 379–391.

117. CROSS, G. L. (1936). The structure of the growing point and the development of the bud scales of *Morus alba* L. *Bull. Torrey bot. Club*, **63**, 451–465.

118. CROSS, G. L. (1937). The origin and development of the foliage leaves and stipules of *Morus alba. Bull. Torrey bot. Club*, **64**, 145–163.

119. CROSS, G. L. and JOHNSTON, T. J. (1941). Structural features of the shoot apices of diploid and colchicine-induced, tetraploid strains of *Vinca rosea* L. *Bull. Torrey bot. Club*, **68**, 618–635.

120. CUSICK, F. (1956). Studies of floral morphogenesis. I. Median bisections of flower primordia in *Primula bulleyana* Forrest. *Trans. R. Soc. Edinb.*, **63**, 153–166.

121. CUSICK, F. (1966). On phylogenetic and ontogenetic fusions. In *Trends in Plant Morphogenesis*, CUTTER, E. G., *et al.*, 170–183. Longmans, London.

122. CUTTER, E. G. (1957). Studies of morphogenesis in the Nymphaeaceae. I. Introduction: some aspects of the morphology of *Nuphar lutea* (L.) Sm. and *Nymphaea alba* L. *Phytomorphology*, **7**, 45–56.

123. CUTTER, E. G. (1959). On a theory of phyllotaxis and histogenesis. *Biol. Rev.*, **34**, 243–263.

124. CUTTER, E. G. (1961). The inception and distribution of flowers in the Nymphaeaceae. *Proc. Linn. Soc. Lond.*, **172**, 93–100.

125. CUTTER, E. G. (1965). Recent experimental studies of the shoot apex and shoot morphogenesis. *Bot. Rev.*, **31**, 7–113.

126. CUTTER, E. G. (1967). Morphogenesis and developmental potentialities of unequal buds. *Phytomorphology*, **17**, 437–445.

127. CUTTER, E. G. (1969). *Plant Anatomy: Experiment and Interpretation.* Part 1. *Cells and Tissues.* Edward Arnold, London.

128. CUTTER, E. G. and FELDMAN, L. J. (1970a). Trichoblasts in *Hydrocharis*. I. Origin, differentiation, dimensions and growth. *Am. J. Bot.*, **57**, 190–201.

129. CUTTER, E. G. and FELDMAN, L. J. (1970b). Trichoblasts in *Hydrocharis*. II. Nucleic acids, proteins and a consideration of cell growth in relation to endopolyploidy. *Am. J. Bot.*, **57**, 202–211.

130. DALE, J. E. (1965). Leaf growth in *Phaseolus vulgaris*. 2. Temperature effects and the light factor. *Ann. Bot.*, *N.S.*, **29**, 293–308.

131. DALE, J. E. (1968). Cell growth in expanding primary leaves of *Phaseolus*. *J. exp. Bot.*, **19**, 322–332.

132. DAVIS, E. L. (1961). Medullary bundles in the genus *Dahlia* and their possible origin. *Am. J. Bot.*, **48**, 108–113.

133. DAVIS, G. J. (1967). *Proserpinaca:* photoperiodic and chemical differentiation of leaf development and flowering. *Pl. Physiol.*, *Lancaster*, **42**, 667–668.

134. DAVIS, G. L. (1966). *Systematic Embryology of the Angiosperms.* Wiley, New York.

135. DENNE, M. P. (1959). Leaf development in *Narcissus pseudonarcissus* L. I. The stem apex. *Ann. Bot.*, *N.S.*, **23**, 121–129.

136. DENNE, M. P. (1960). Leaf development in *Narcissus pseudonarcissus* L. II. The comparative development of scale and foliage leaves. *Ann. Bot.*, *N.S.*, **24**, 32–47.

137. DENNE, M. P. (1966a). Morphological changes in the shoot apex of *Trifolium repens* L. I. Changes in the vegetative apex during the plastochron. *N. Z. Jl Bot.*, **4**, 300–314.

138. DENNE, M. P. (1966b). Diurnal and plastochronic changes in the shoot apex of *Tradescantia fluminensis* Vell. *N. Z. Jl Bot.*, **4**, 444–454.

139. DENNE, M. P. (1966c). Leaf development in *Trifolium repens*. *Bot. Gaz.*, **127**, 202–210.

140. DE VRIES, H. A. M. A. (1968a). Development of the structure of the normal smooth cuticle of the apple 'Golden Delicious'. *Acta bot. neerl.*, **17**, 229–241.

141. DE VRIES, H. A. M. A. (1968b). Development of the structure of the russeted apple skin. *Acta bot. neerl.*, **17**, 405–415.

142. DIGBY, J. and WANGERMANN, E. (1965). A note on the effect of the shoot and root apex on secondary thickening in pea radicles. *New Phytol.*, **64**, 168–170.

143. DIGBY, J. and WAREING, P. F. (1966a). The effect of applied growth hormones on cambial division and the differentiation of the cambial derivatives. *Ann. Bot.*, *N.S.*, **30**, 539–548.

144. DIGBY, J. and WAREING, P. F. (1966b). The relationship between endogenous hormone levels in the plant and seasonal aspects of cambial activity. *Ann. Bot.*, *N.S.*, **30**, 607–622.

145. DIGBY, J. and WAREING, P. F. (1966c). The effect of growth hormones on cell division and expansion in liquid suspension cultures of *Acer pseudoplatanus*. *J. exp. Bot.*, **17**, 718–725.

146. DITTMER, H. J. and TALLEY, B. P. (1964). Gross morphology of tap roots of desert cucurbits. *Bot. Gaz.*, **125**, 121–126.

147. DOBBINS, D. R. (1969). Studies on the anomalous cambial activity in *Doxantha unguis-cati* (Bignoniaceae). I. Development of the vascular pattern. *Can. J. Bot.*, **47**, 2101–2106.

148. DOBBINS, D. R. (1971). *Studies on anomalous cambial activity in* Doxantha unguis-cati *(Bignoniaceae)*. Ph. D. dissertation, Univ. of Massachusetts.

149. DORE, J. (1965). Physiology of regeneration in cormophytes. *Handb. PflPhysiol.*, **15**, 2, 1–91.

150. DOSTÁL, R. (1967). *On Integration in Plants*. Harvard University Press, Cambridge, Mass.

151. DOWNTON, W. J. S. and TREGUNNA, E. B. (1968). Carbon dioxide compensation—its relation to photosynthetic carboxylation reactions, systematics of the Gramineae, and leaf anatomy. *Can. J. Bot.*, **46**, 207–215.

152. DYCUS, A. M. and KNUDSON, L. (1957). The role of the velamen of the aerial roots of orchids. *Bot. Gaz.*, **119**, 78–87.

153. EAGLES, C. F. and WAREING, P. F. (1963). Dormancy regulators in woody plants. Experimental induction of dormancy in *Betula pubescens*. *Nature, Lond.*, **199**, 874–875.

154. EAMES, A. J. (1961). *Morphology of the Angiosperms*. McGraw-Hill, New York.

155. ECHLIN, P. (1968). Pollen. *Scient. Am.*, **218**, 4, 81–90.

156. ECHLIN, P. and GODWIN, H. (1968a). The ultrastructure and ontogeny of pollen in *Helleborus foetidus* L. I. The development of the tapetum and Ubisch bodies. *J. Cell Sci.*, **3**, 161–174.

157. ECHLIN, P. and GODWIN, H. (1968b). The ultrastructure and ontogeny of pollen in *Helleborus foetidus* L. II. Pollen grain development through the callose special wall stage. *J. Cell Sci.*, **3**, 175–186.

158. ECHLIN, P. and GODWIN, H. (1969). The ultrastructure and ontogeny of pollen in *Helleborus foetidus* L. III. The formation of the pollen grain wall. *J. Cell Sci.*, **5**, 459–477.

159. ECKARDT, T. (1941). Kritische Untersuchungen über das primäre Dickenwachstum bei Monokotylen, mit Ausblick auf dessen Verhältnis zur sekundären Verdickung. *Bot. Arch.*, **42**, 289–334.

160. EDWARDS, M. M. (1968a). Dormancy in seeds of charlock. I. Developmental anatomy of the seed. *J. exp. Bot.*, **19**, 575–582.

161. EDWARDS, M. M. (1968b). Dormancy in seeds of charlock. II. The influence of the seed coat. *J. exp. Bot.*, **19**, 583–600.

162. EDWARDS, M. M. (1968c). Dormancy in seeds of charlock. III. Occurrence and mode of action of an inhibitor associated with dormancy. *J. exp. Bot.*, **19**, 601–610.

163. ERICKSON, R. O. and MICHELINI, F. J. (1957). The plastochron index. *Am. J. Bot.*, **44**, 297–305.

164. ESAU, K. (1946). Morphology of reproduction in guayule and certain other species of *Parthenium*. *Hilgardia*, **17**, 61–120.

165. ESAU, K. (1954). Primary vascular differentiation in plants. *Biol. Rev.*, **29**, 46–86.

166. ESAU, K. (1964). Structure and development of the bark in dicotyledons. In *The Formation of Wood in Forest Trees*, ZIMMERMANN, M. H., 37–50. Academic Press, New York.

167. ESAU, K. (1965a). *Plant Anatomy*. 2nd edition. Wiley, New York.

168. ESAU, K. (1965b). *Vascular Differentiation in Plants*. Holt, Rinehart and Winston, New York.

169. ESAU, K. (1967). Minor veins in *Beta* leaves: structure related to function. *Proc. Am. phil. Soc.*, **111**, 219–233.

170. ESAU, K. and CHEADLE, V. I. (1969). Secondary growth in *Bougainvillea*. *Ann. Bot.*, *N.S.*, **33**, 807–819.

171. EVANS, L. J., Editor (1969). *The Induction of Flowering. Some Case Histories*. Cornell Univ. Press, Ithaca, New York.

172. EVANS, P. S. (1965). Intercalary growth in the aerial shoot of *Eleocharis acuta* R. Br. Prodr. I. Structure of the growing zone. *Ann. Bot.*, *N.S.*, **29**, 205–217.

173. EVANS, P. S. (1969). Intercalary growth in the aerial shoot of *Eleocharis acuta* R. Br. II. Development of the main internode. *N.Z. Jl Bot.*, **7**, 36–42.

174. FAHN, A. and ZOHARY, M. (1955). On the pericarpal structure of the legumen, its evolution and relation to dehiscence. *Phytomorphology*, **5**, 99–111.

175. FELDMAN, L. J. and CUTTER, E. G. (1970a). Regulation of leaf form in *Centaurea solstitialis* L. I. Leaf development on whole plants in sterile culture. *Bot. Gaz.*, **131**, 31–39.

176. FELDMAN, L. J. and CUTTER, E. G. (1970b). Regulation of leaf form in *Centaurea solstitialis* L. II. The developmental potentialities of excised leaf primordia in sterile culture. *Bot. Gaz.*, **131**, 39–49.

177. FISHER, J. B. (1970a). Development of the intercalary meristem of *Cyperus alternifolius*. *Am. J. Bot.*, **57**, 691–703.

178. FISHER, J. B. (1970b). Control of the internodal intercalary meristem of *Cyperus alternifolius*. *Am. J. Bot.*, **57**, 1017–1026.

179. FOARD, D. E., HABER, A. H. and FISHMAN, T. N. (1965). Initiation of lateral root primordia without completion of mitosis and without cytokinesis in uniseriate pericycle. *Am. J. Bot.*, **52**, 580–590.

180. FORDE, B. J. (1966). Effect of various environments on the anatomy and growth of perennial ryegrass and cocksfoot. I. Leaf growth. *N.Z. Jl Bot.*, **4**, 455–468.

181. FORMAN, M. and JENSEN, W. A. (1965). Respiration and embryogenesis in cotton. *Pl. Physiol., Lancaster*, **40**, 765–769.

182. FOSKET, D. E. and MIKSCHE, J. P. (1966). A histochemical study of the seedling shoot apical meristem of *Pinus lambertiana*. *Am. J. Bot.*, **53**, 694–702.

183. FOSSARD, R. A. DE (1969). Development and histochemistry of the endothecium in the anthers of *in vitro* grown *Chenopodium rubrum* L. *Bot. Gaz.*, **130**, 10–22.

184. FOSTER, A. S. (1928). Salient features of the problem of bud-scale morphology. *Biol. Rev.*, **3**, 123–164.
185. FOSTER, A. S. (1929). Investigations on the morphology and comparative history of development of foliar organs. I. The foliage leaves and cataphyllary structures in the horsechestnut (*Aesculus hippocastanum* L.). *Am. J. Bot.*, **16**, 441–501.
186. FOSTER, A. S. (1935a). A histogenetic study of foliar determination in *Carya Buckleyi* var. *arkansana*. *Am. J. Bot.*, **22**, 88–147.
187. FOSTER, A. S. (1935b). Comparative histogenesis of foliar transition forms in *Carya*. *Univ. Calif. Publs Bot.*, **19**, 159–185.
188. FOSTER, A. S. (1936). Leaf differentiation in angiosperms. *Bot. Rev.*, **2**, 349–372.
189. FOSTER, A. S. (1938). Structure and growth of the shoot apex in *Ginkgo biloba*. *Bull. Torrey bot. Club*, **65**, 531–556.
190. FOSTER, A. S. (1939). Problems of structure, growth and evolution in the shoot apex of seed plants. *Bot. Rev.*, **5**, 454–470.
191. FOSTER, A. S. (1952). Foliar venation in angiosperms from an ontogenetic standpoint. *Am. J. Bot.*, **39**, 752–766.
192. FOSTER, A. S. (1961). The phylogenetic significance of dichotomous venation in angiosperms. In *Recent Advances in Botany*, 971–975. Univ. Toronto Press.
193. FOSTER, A. S. (1966). Morphology of anastomoses in the dichotomous venation of *Circaeaster*. *Am. J. Bot.*, **53**, 588–599.
194. FOSTER, A. S. and ARNOTT, H. J. (1960). Morphology and dichotomous vasculature of the leaf of *Kingdonia uniflora*. *Am. J. Bot.*, **47**, 684–698.
195. DE LA FUENTE, R. K. and LEOPOLD, A. C. (1968). Senescence processes in leaf abscission. *Pl. Physiol., Lancaster*, **43**, 1496–1502.
196. GALUN, E., JUNG, Y. and LANG, A. (1963). Morphogenesis of floral buds of cucumber cultured *in vitro*. *Devl Biol.*, **6**, 370–387.
197. GARNER, W. W. and ALLARD, H. A. (1923). Further studies in photoperiodism, the response of the plant to relative length of day and night. *J. agric. Res.*, **23**, 871–920.
198. GARRISON, R. and WETMORE, R. H. (1961). Studies in shoot-tip abortion: *Syringa vulgaris*. *Am. J. Bot.*, **48**, 789–795.
199. GAUDET, J. J. (1964). Morphology of *Marsilea vestita*. II. Morphology of the adult land and submerged leaves. *Am. J. Bot.*, **51**, 591–597.
200. GAUDET, J. J. (1965a). The effect of various environmental factors on the leaf form of the aquatic fern *Marsilea vestita*. *Physiologia Pl.*, **18**, 674–686.
201. GAUDET, J. J. (1965b). Morphology of *Marsilea vestita*. III. Morphogenesis of the leaves of etiolated plants. *Am. J. Bot.*, **52**, 716–719.
202. GEIGER, D. R. and CATALDO, D. A. (1969). Leaf structure and translocation in sugar beet. *Pl. Physiol., Lancaster*, **44**, 45–54.
202a. GIBBONS, G. S. B. and WILKINS, M. B. (1970). Growth inhibitor production by root caps in relation to geotropic responses. *Nature, Lond.*, **226**, 558–559.
203. GIFFORD, E. M., Jr. (1954). The shoot apex in angiosperms. *Bot. Rev.*, **20**, 477–529.
204. GIFFORD, E. M., Jr. (1964). Developmental studies of vegetative and floral meristems. *Brookhaven Symp. Biol.*, **16**, 126–137.
205. GIFFORD, E. M., Jr. (1969). Initiation and early development of the

inflorescence in pineapple (*Ananas comosus*, var. Smooth Cayenne treated with acetylene. *Am. J. Bot.*, **56**, 892–897.

206. GIFFORD, E. M., Jr., KUPILA, S. and YAMAGUCHI, S. (1963). Experiments in the application of H^3-thymidine and adenine-8-C^{14} to shoot tips. *Phytomorphology*, **13**, 14–22.

207. GIFFORD, E. M., Jr. and STEWART, K. D. (1965). Ultrastructure of vegetative and reproductive apices of *Chenopodium album*. *Science, N.Y.*, **149**, 75–77.

208. GIFFORD, E. M., Jr. and STEWART, K. D. (1967). Ultrastructure of the shoot apex of *Chenopodium album* and certain other seed plants. *J. Cell Biol.*, **33**, 131–142.

209. GIFFORD, E. M., Jr. and TEPPER, H. B. (1961). Ontogeny of the inflorescence in *Chenopodium album*. *Am. J. Bot.*, **48**, 657–667.

210. GIFFORD, E. M., Jr. and TEPPER, H. B. (1962a). Histochemical and autoradiographic studies of floral induction in *Chenopodium album*. *Am. J. Bot.*, **49**, 706–714.

211. GIFFORD, E. M., Jr. and TEPPER, H. B. (1962b). Ontogenetic and histochemical changes in the vegetative shoot tip of *Chenopodium album*. *Am. J. Bot.*, **49**, 902–911.

212. GIROLAMI, G. (1953). Relation between phyllotaxis and primary vascular organization in *Linum*. *Am. J. Bot.*, **40**, 618–625.

213. GLATER, R. A. B., SALBERG, R. A. and SCOTT, F. M. (1962). A developmental study of the leaves of *Nicotiana glutinosa* as related to their smog sensitivity. *Am. J. Bot.*, **49**, 954–970.

214. GODWIN, H. (1968). The origin of the exine. *New Phytol.*, **67**, 667–676.

215. GOEBEL, K. (1880). Beiträge zur Morphologie und Physiologie des Blattes. *Bot. Ztg*, **38**, 753–760, 769–778, 785–795, 801–815, 817–826, 833–845.

216. GOEBEL, K. (1884). Vergleichende Entwicklungsgeschichte der Pflanzenorgane. Schenk's *Handbuch der Botanik*, **3**, 99–432.

217. GOODWIN, R. H. and AVERS, C. J. (1956). Studies on roots. III. An analysis of root growth in *Phleum pratense* using photomicrographic records. *Am. J. Bot.*, **43**, 479–487.

218. GRÉGOIRE, V. (1938). La morphogénèse et l'autonomie morphologique de l'appareil floral. I. Le carpelle. *Cellule*, **17**, 287–452.

219. GREYSON, R. I. and TEPFER, S. S. (1966). An analysis of stamen filament growth of *Nigella hispanica*. *Am. J. Bot.*, **53**, 485–490.

220. GREYSON, R. I. and TEPFER, S. S. (1967). Emasculation effects on the stamen filament of *Nigella hispanica* and their partial reversal by gibberellic acid. *Am. J. Bot.*, **54**, 971–976.

221. GRIFFITHS, H. J. and AUDUS, L. J. (1964). Organelle distribution in the statocyte cells of the root-tip of *Vicia faba* in relation to geotropic stimulation. *New Phytol.*, **63**, 319–333.

222. GUHA, S. and MAHESHWARI, S. C. (1964). *In vitro* production of embryos from anthers of *Datura*. *Nature, Lond.*, **204**, 497.

223. GUHA, S. and MAHESHWARI, S. C. (1966). Cell division and differentiation of embryos in the pollen grains of *Datura in vitro*. *Nature, Lond.*, **212**, 97–98.

224. GUNCKEL, J. E. and WETMORE, R. H. (1946). Studies of development in long shoots and short shoots of *Ginkgo biloba* L. I. The origin and pattern of development of the cortex, pith and procambium. *Am. J. Bot.*, **33**, 285–295.

225. GUNNING, B. E. S., PATE, J. S. and BRIARTY, L. G. (1968). Specialized 'transfer cells' in minor veins of leaves and their possible significance in phloem translocation. *J. Cell Biol.*, **37**, C7–C12.

225a. GUNNING, B. E. S., PATE, J. S. and GREEN, L. W. (1971). Transfer cells in the vascular system of stems: taxonomy, association with nodes, and structure. *Protoplasma*, **71**, in press.

226. GUSTAFSON, F. G. (1961). Development of fruits. *Handb. PflPhysiol.*, **14**, 951–958.

227. GUTTENBERG, H. VON (1947). Studien über die Entwicklung des Wurzel-vegetationspunktes der Dikotyledonen. *Planta*, **35**, 360–396.

228. GUTTENBERG, H. VON (1960). *Grundzüge der Histogenese höherer Pflanzen.* I. *Die Angiospermen.* Gebruder Borntraeger, Berlin.

229. GUTTENBERG, H. VON (1964). Die Entwicklung der Wurzel. *Phyto-morphology*, **14**, 265–287.

230. HABER, A. H. (1962). Nonessentiality of concurrent cell divisions for degree of polarization of leaf growth. I. Studies with radiation-induced mitotic inhibition. *Am. J. Bot.*, **49**, 583–589.

231. HABERLANDT, G. (1914). *Physiological Plant Anatomy.* Macmillan, London.

232. HACCIUS, B. (1965a). Weitere Untersuchungen über Somatogenese aus den Suspensorenzellen von *Eranthis hiemalis*-Embryonen. *Planta*, **64**, 219–224.

233. HACCIUS, B. (1965b). Haben 'Gewebekultur-Embryonen' einen Suspensor? *Ber. dt. bot. Ges.*, **78**, 11–21.

234. HACKETT, W. P. and SACHS, R. M. (1968). Experimental separation of inflorescence development from initiation in *Bougainvillea. Proc. Am. Soc. hort. Sci.*, **92**, 615–621.

235. HAIGH, W. G. and GUARD, A. T. (1963). Distribution of mitoses in root tip of *Zea mays. Bot. Gaz.*, **124**, 421–423.

236. HALPERIN, W. (1966). Alternative morphogenetic events in cell suspensions. *Am. J. Bot.*, **53**, 443–453.

237. HALPERIN, W. and JENSEN, W. A. (1967). Ultrastructural changes during growth and embryogenesis in carrot cell cultures. *J. Ultrastruct. Res.*, **18**, 428–443.

238. HANDA, T. (1936). Abnormal vascular bundle in the stem of *Campsis grandiflora* K. Schum. *Jap. J. Bot.*, **8**, 47–58.

239. HANNAM, R. V. (1968). Leaf growth and development in the young tobacco plant. *Aust. J. biol. Sci.*, **21**, 855–870.

240. HANSTEIN, J. (1868). Die Scheitelzellgruppe im Vegetationspunkt der Phanerogamen. *Festschr. Niederrhein. Gesell. Natur- und Heilkunde.* 1868: 109–134.

241. HARA, N. (1957). On the types of the marginal growth in dicotyledonous foliage leaves. *Bot. Mag., Tokyo*, **70**, 108–114.

242. HARRIS, W. M. and SPURR, A. R. (1969). Chromoplasts of tomato fruits. I. Ultrastructure of low-pigment and high-beta mutants. Carotene analyses. *Am. J. Bot.*, **56**, 369–379.

243. HATCH, M. D. and SLACK, C. R. (1966). Photosynthesis by sugar-cane leaves. A new carboxylation reaction and the pathway of sugar formation. *Biochem. J.*, **101**, 103–111.

244. HAYWARD, H. E. (1938). *The Structure of Economic Plants.* Macmillan, New York.

245. HEIMSCH, C. (1951). Development of vascular tissues in barley roots. *Am. J. Bot.*, **38**, 523–537.

246. HERTEL, R., DE LA FUENTE, R. K., and LEOPOLD, A. C. (1969). Geotropism and the lateral transport of auxin in the corn mutant amylomaize. *Planta*, **88**, 204–214.

247. HESLOP-HARRISON, J. (1959). Growth substances and flower morphogenesis. *J. Linn. Soc. (Bot.)*, **56**, 269–281.

248. HESLOP-HARRISON, J. (1963). Ultrastructural aspects of differentiation in sporogenous tissue. *Symp. Soc. exp. Biol.*, **17**, 315–340.

249. HESLOP-HARRISON, J. (1964). Sex expression in flowering plants. *Brookhaven Symp. Biol.*, **16**, 109–125.

250. HESLOP-HARRISON, J. (1968a). Tapetal origin of pollen-coat substances in *Lilium*. *New Phytol.*, **67**, 779–786.

251. HESLOP-HARRISON, J. (1968b). Pollen wall development. *Science, N.Y.*, **161**, 230–237.

252. HESLOP-HARRISON, J. (1968c). Synchronous pollen mitosis and the formation of the generative cell in massulate orchids. *J. Cell Sci.*, **3**, 457–466.

253. HESLOP-HARRISON, J. and DICKINSON, H. G. (1969). Time relationships of sporopollenin synthesis associated with tapetum and microspores in *Lilium*. *Planta*, **84**, 199–214.

254. HESLOP-HARRISON, J. and HESLOP-HARRISON, Y. (1969). *Cannabis sativa* L. In *The Induction of Flowering*, EVANS, L. T., 205–226. Cornell Univ. Press, Ithaca, New York.

255. HESLOP-HARRISON, J. and MACKENZIE, A. (1967). Autoradiography of soluble (2-^{14}C) thymidine derivatives during meiosis and microsporogenesis in *Lilium* anthers. *J. Cell Sci.*, **2**, 387–400.

256. HEYWOOD, V. H. (1968). Scanning electron microscopy and microcharacters in the fruits of the Umbelliferae–Caucalideae. *Proc. Linn. Soc. Lond.*, **179**, 287–289.

257. HILLMAN, W. S. (1962). *The Physiology of Flowering*. Holt, Rinehart and Winston, New York.

258. HITCH, P. A. and SHARMAN, B. C. (1968). Initiation of procambial strands in leaf primordia of *Dactylis glomerata* L. as an example of a temperate herbage grass. *Ann. Bot., N.S.*, **32**, 153–164.

259. HODGE, A. J., MCLEAN, J. D. and MERCER, F. V. (1955). Ultrastructure of the lamellae and grana in the chloroplasts of *Zea mays* L. *J. biophys. biochem. Cytol.*, **1**, 605–614.

260. HOEFERT, L. L. (1969). Ultrastructure of *Beta* pollen. I. Cytoplasmic constituents. *Am. J. Bot.*, **56**, 363–368.

261. HOUGHTALING, H. B. (1935). A developmental analysis of size and shape in tomato fruits. *Bull. Torrey bot. Club*, **62**, 243–251.

262. HUGHES, A. P. (1959). Effects of the environment on leaf development in *Impatiens parviflora* DC. *J. Linn. Soc. (Bot.)*, **56**, 161–165.

263. HUSSEY, G. (1963). Growth and development in the young tomato. I. The effect of temperature and light intensity on growth of the shoot apex and leaf primordia. *J. exp. Bot.*, **14**, 316–325.

263a. HYDE, E. O. C. (1954). The function of the hilum in some Papilionaceae in relation to the ripening of the seed and the permeability of the testa. *Ann. Bot., N.S.*, **18**, 241–256.

264. INAMDAR, J. A. (1968a). Epidermal structure and ontogeny of stomata in some Verbenaceae. *Ann. Bot., N.S.*, **33**, 55–66.

265. INAMDAR, J. A. (1968b). Ontogeny of stomata in some Oleaceae. *Proc. Indian Acad. Sci.*, **67**, 157–164.
266. INAMDAR, J. A. (1968c). Development of stomata in vegetative and floral organs in some Caryophyllaceae. *Aust. J. Bot.*, **16**, 445–449.
267. INAMDAR, J. A., GOPAL, B. V. and CHOHAM, A. J. (1968). Development of normal and abnormal stomata in some Araliaceae. *Ann. Bot.*, *N.S.*, **33**, 67–73.
268. IWAHORI, S. and LYONS, J. M. (1969). Accelerating tomato fruit maturity with ethrel. *Calif. Agric.*, **23**, 6, 17–18.
269. JACOBS, W. P. (1947). The development of the gynophore of the peanut plant, *Arachis hypogaea* L. I. The distribution of mitoses, the region of greatest elongation, and the maintenance of vascular continuity in the intercalary meristem. *Am. J. Bot.*, **34**, 361–370.
270. JACOBS, W. P. (1968). Hormonal regulation of leaf abscission. *Pl. Physiol.*, *Lancaster*, **43**, 1480–1495.
271. JACOBS, W. P. and MORROW, I. B. (1957). A quantitative study of xylem development in the vegetative shoot apex of *Coleus*. *Am. J. Bot.*, **44**, 823–842.
272. JACOBS, W. P. and MORROW, I. B. (1961). A quantitative study of mitotic figures in relation to development in the apical meristem of vegetative shoots of *Coleus*. *Devl Biol.*, **3**, 569–587.
273. JANCZEWSKI, E. DE (1874). Recherches sur l'accroissement terminal des racines dans les phanérogames. *Annls Sci. nat. (Bot.)*, Sér. **5**, **20**, 162–201.
274. JANE, F. W. (1962). *The Structure of Wood*. A. and C. Black, London.
275. JENSEN, G. E. and VALDOVINOS, J. G. (1967). Fine structure of abscission zones. I. Abscission zones of the pedicels of tobacco and tomato flowers at anthesis. *Planta*, **77**, 298–318.
276. JENSEN, W. A. (1955). The histochemical localization of peroxidase in roots and its induction by indoleacetic acid. *Pl. Physiol.*, *Lancaster*, **30**, 426–432.
277. JENSEN, W. A. (1958). The nucleic acid and protein content of root tip cells of *Vicia faba* and *Allium cepa*. *Expl Cell Res.*, **14**, 575–583.
278. JENSEN, W. A. (1964). Cell development during plant embryogenesis. *Brookhaven Symp. Biol.*, **16**, 179–202.
279. JENSEN, W. A. (1968). Cotton embryogenesis: the tube-containing endoplasmic reticulum. *J. Ultrastruct. Res.*, **22**, 296–302.
280. JENSEN, W. A. (1969). Cotton embryogenesis: pollen tube development in the nucellus. *Can. J. Bot.*, **47**, 383–385.
281. JENSEN, W. A. and FISHER, D. B. (1967). Cotton embryogenesis: double fertilization. *Phytomorphology*, **17**, 261–269.
282. JENSEN, W. A. and FISHER, D. B. (1968). Cotton embryogenesis: the sperm. *Protoplasma*, **65**, 277–286.
283. JENSEN, W. A. and FISHER, D. B. (1969). Cotton embryogenesis: the tissues of the stigma and style and their relationship to the pollen tube. *Planta*, **84**, 97–121.
284. JENSEN, W. A. and KAVALJIAN, L. G. (1958). An analysis of cell morphology and the periodicity of division in the root tip of *Allium cepa*. *Am. J. Bot.*, **45**, 365–372.
285. JOHANSEN, D. A. (1950). *Plant Embryology*. Chronica Botanica Co., Waltham, Mass.

286. JOHNSON, M. A. and TRUSCOTT, F. H. (1956). On the anatomy of *Serjania*. I. Path of the bundles. *Am. J. Bot.*, **43**, 509–518.

287. JONES, H. (1955a). Heterophylly in some species of *Callitriche* with especial reference to *Callitriche intermedia*. *Ann. Bot.*, *N.S.*, **19**, 225–245.

288. JONES, H. (1955b). Further studies on heterophylly in *Callitriche intermedia*: leaf development and experimental induction of ovate leaves. *Ann. Bot.*, *N.S.*, **19**, 369–388.

289. JONES, H. (1956). Morphological aspects of leaf expansion, especially in relation to changes in leaf form. In *The Growth of Leaves*, MILTHORPE, F. L., 93–105. Butterworths Scientific Publications, London.

290. JONES, R. L. (1969a). The fine structure of barley aleurone cells. *Planta*, **85**, 359–375.

291. JONES, R. L. (1969b). Gibberellic acid and the fine structure of barley aleurone cells. I. Changes during the lag-phase of α-amylase synthesis. *Planta*, **87**, 119–133.

292. JONES, R. L. (1969c). Gibberellic acid and the fine structure of barley aleurone cells. II. Changes during the synthesis and secretion of α-amylase. *Planta*, **88**, 73–86.

293. JONES, R. L. and PHILLIPS, I. D. J. (1966). Organs of gibberellin synthesis in light-grown sunflower plants. *Pl. Physiol.*, *Lancaster*, **41**, 1381–1386.

294. JOSHI, P. C. and BALL, E. (1968a). Growth of isolated palisade cells of *Arachis hypogaea in vitro. Devl Biol.*, **17**, 308–325.

295. JOSHI, P. C. and BALL, E. (1968b). Growth values and regenerative potentiality in mesophyll cultures of *Arachis hypogaea. Z. PflPhysiol.*, **59**, 109–123.

296. JOST, L. (1893). Über Beziehungen zwischen der Blattentwickelung und der Gefässbildung in den Pflanze. *Bot. Ztg*, **51**, 89–138.

297. JUNIPER, B. E. and CLOWES, F. A. L. (1965). Cytoplasmic organelles and cell growth in root caps. *Nature, Lond.*, **208**, 864–865.

298. JUNIPER, B. E., GROVES, S., LANDAU-SCHACHAR, B. and AUDUS, L. J. (1966). Root cap and the perception of gravity. *Nature, Lond.*, **209**, 93–94.

299. KANTA, K., RANGASWAMY, N. S. and MAHESHWARI, P. (1962). Test-tube fertilization in a flowering plant. *Nature, Lond.*, **194**, 1214–1217.

300. KAPLAN, D. R. (1967). Floral morphology, organogenesis and interpretation of the inferior ovary in *Downingia bacigalupii. Am. J. Bot.*, **54**, 1274–1290.

301. KAPLAN, D. R. (1968a). Structure and development of the perianth in *Downingia bacigalupii. Am .J. Bot.*, **55**, 406–420.

302. KAPLAN, D. R. (1968b). Histogenesis of the androecium and gynoecium in *Downingia bacigalupii. Am. J. Bot.*, **55**, 933–950.

303. KAPLAN, D. R. (1969). Sporogenesis and gametogenesis in *Downingia* (Campanulaceae; Lobelioideae). *Bull. Torrey bot. Club*, **96**, 418–434.

304. KAPLAN, D. R. (1970). Comparative foliar histogenesis in *Acorus calamus* and its bearing on the phyllode theory of monocotyledonous leaves. *Am. J. Bot.*, **57**, 331–361.

305. KAUFMAN, P. B. (1959). Development of the shoot of *Oryza sativa* L. II. Leaf histogenesis. *Phytomorphology*, **9**, 277–311.

306. KAUFMAN, P. B., CASSELL, S. J. and ADAMS, P. A. (1965). On nature of intercalary growth and cellular differentiation in internodes of *Avena sativa. Bot. Gaz.*, **126**, 1–13.

307. KAUFMAN, P. B., PETERING, L. B., YOCUM, C. S. and BAIC, D. (1970). Ultrastructural studies on stomata development in internodes of *Avena sativa. Am. J. Bot.*, **57**, 33–49.

308. KENDE, H. (1965). Kinetinlike factors in the root exudate of sunflowers. *Proc. natn. Acad. Sci. U.S.A.*, **53**, 1302–1307.

309. KENNEDY, R. W. and FARRAR, J. L. (1965a). Tracheid development in tilted seedlings. In *Cellular Ultrastructure of Woody Plants*, CÔTÉ, W. A., Jr., 419–453. Syracuse University Press.

310. KENNEDY, R. W. and FARRAR, J. L. (1965b). Induction of tension wood with the anti-auxin 2,3,5-tri-iodobenzoic acid. *Nature, Lond.*, **208**, 406–407.

311. KINET, J.-M. (1966). Action du 2-thiouracile sur l'évolution du méristème caulinaire de *Sinapis alba* L. au cour du passage de l'état végétatif à l'état floral. *Les Congrès et Colloques de l'Université de Liège*, **38**, Les Phytohormones et l'Organogenèse, 243–263.

312. KING, E. E. and LANE, H. C. (1969). Abscission of cotton flower buds and petioles caused by protein from boll weevil larvae. *Pl. Physiol., Lancaster*, **44**, 903–906.

313. KNOX, R. B. and EVANS, L. T. (1966). Inflorescence initiation in *Lolium temulentum* L. VIII. Histochemical changes at the shoot apex during induction. *Aust. J. biol. Sci.*, **19**, 233–245.

314. KNOX, R. B. and EVANS, L. T. (1968). Inflorescence initiation in *Lolium temulentum* L. XII. An autoradiographic study of evocation in the shoot apex. *Aust. J. biol. Sci.*, **21**, 1083–1094.

315. KNOX, R. B. and HESLOP-HARRISON, J. (1969). Cytochemical localization of enzymes in the wall of the pollen grain. *Nature, Lond.*, **223**, 92–94.

316. KONAR, R. N. and NATARAJA, K. (1965). Experimental studies in *Ranunculus sceleratus* L. Development of embryos from the stem epidermis. *Phytomorphology*, **15**, 132–137.

317. KONINGS, H. (1967). On the mechanism of the transverse distribution of auxin in geotropically exposed pea roots. *Acta bot. neerl.*, **16**, 161–176.

318. KONINGS, H. (1968). The significance of the root cap for geotropism. *Acta bot. neerl.*, **17**, 203–211.

319. LAETSCH, W. M. (1968). Chloroplast specialization in dicotyledons possessing the C_4-dicarboxylic acid pathway of photosynthetic CO_2 fixation. *Am. J. Bot.*, **55**, 875–883.

319a. LAETSCH, W. M. (1969). Relationship between chloroplast structure and photosynthetic carbon-fixation pathways. *Sci. Prog., Oxf.*, **57**, 323–351.

320. LAETSCH, W. M. and PRICE, I. (1969). Development of the dimorphic chloroplasts of sugar cane. *Am. J. Bot.*, **56**, 77–87.

321. LAETSCH, W. M., STETLER, D. A. and VLITOS, A. J. (1966). The ultrastructure of sugar cane chloroplasts. *Z. PflPhysiol.*, **54**, 472–474.

322. LANCE, A. (1957a). Recherches cytologiques sur l'évolution de quelques méristèmes apicaux et sur ses variations provoquées par des traitements photopériodiques. *Annls Sci. nat. (Bot.)*, Sér. 11, **18**, 91–422.

323. LANCE, A. (1957b). Sur l'infrastructure des cellules apicales de '*Chrysanthemum segetum*' L. (Composées). *C. r. hebd. Séanc. Acad. Sci., Paris*, **245**, 352–355.

324. LANCE-NOUGARÈDE, A. (1961). Sur l'incorporation de l'adénine marquée au tritium (^{3}H) dans les noyaux et le cytoplasme des cellules de deux méristèmes caulinaires: *Lupinus albus* (Papilionacées) et *Teucrium scorodonia* (Labiées). *C. r. hebd. Séanc. Acad. Sci.*, Paris, **252**, 1504–1506.

325. LARSON, P. R. (1962). The indirect effect of photoperiod on tracheid diameter in *Pinus resinosa. Am. J. Bot.*, **49**, 132–137.

326. LARSON, P. R. (1964). Some indirect effects of environment on wood formation. In *The Formation of Wood in Forest Trees*, ZIMMERMANN, M. H., 345–365. Academic Press, New York.

327. LAWTON, J. R. S. and LAWTON, J. R. (1967). The morphology of the dormant embryo and young seedling of five species of *Dioscorea* from Nigeria. *Proc. Linn. Soc. Lond.*, **178**, 153–159.

328. LEACH, R. W. A. and WAREING, P. F. (1967). Distribution of auxin in horizontal woody stems in relation to gravimorphism. *Nature, Lond.*, **214**, 1025–1027.

329. LEECH, J. H., MOLLENHAUER, H. H. and WHALEY, W. G. (1963). Ultrastructural changes in the root apex. *Symp. Soc. exp. Biol.*, **17**, 74–84.

330. LEOPOLD, A. C. (1964). *Plant Growth and Development*. McGraw-Hill, New York.

331. LLOYD, D. G. (1968). Pollen tube growth and seed set in self-incompatible and self-compatible *Leavenworthia* (Cruciferae) populations. *New Phytol.*, **67**, 179–195.

332. LOISEAU, J.-E. (1959). Observations et expérimentation sur la phyllotaxie, et le fonctionnement du sommet végétatif chez quelques Balsaminacées. *Annls Sci. nat. (Bot.)*, Sér. **11**, **20**, 1–214.

333. LOISEAU, J.-E. (1962). Activité mitotique des cellules superficielles du sommet végétatif caulinaire. *Mem. Soc. bot. Fr.* (1962), 14–23.

334. LONGMAN, K. A. (1961). Factors affecting flower initiation in certain conifers. *Proc. Linn. Soc., Lond.*, **172**, 124–127.

335. LOOMIS, R. S. and TORREY, J. G. (1964). Chemical control of vascular cambium initiation in isolated radish roots. *Proc. natn. Acad. Sci. U.S.A.*, **52**, 3–11.

336. LUCKWILL, L. C. (1959). Fruit growth in relation to internal and external chemical stimuli. In *Cell, Organism and Milieu*, RUDNICK, D., 223–251. *17th Symp. Dev. Growth*. Ronald Press, New York.

337. LYNDON, R. F. (1968). Changes in volume and cell number in the different regions of the shoot apex of *Pisum* during a single plastochron. *Ann. Bot.*, *N.S.*, **32**, 371–390.

338. LYNDON, R. F. (1970a). Rates of cell division in the shoot apical meristem of *Pisum. Ann. Bot.*, *N.S.*, **34**, 1–17.

339. LYNDON, R. F. (1970b). Planes of cell division and growth in the shoot apex of *Pisum. Ann. Bot.*, *N.S.*, **34**, 19–28.

340. LYNDON, R. F. (1970c). DNA, RNA and protein in the pea shoot apex in relation to leaf initiation. *J. exp. Bot.*, **21**, 286–291.

341. MACARTHUR, M. and WETMORE, R. H. (1939). Developmental studies in the apple fruit in the varieties McIntosh Red and Wagener. I. Vascular anatomy. *J. Pomol.*, **17**, 218–232.

342. MACARTHUR, M. and WETMORE, R. H. (1941). Developmental studies of the apple fruit in the varieties McIntosh Red and Wagener. II. An analysis of development. *Can. J. Res.*, C, **19**, 371–382.

343. MACLEOD, A. M. (1969). The utilization of cereal seed reserves. *Sci. Prog., Oxf.*, **57**, 99–112.

344. MAHESHWARI, P. (1950). *An Introduction to the Embryology of Angiosperms.* McGraw-Hill, New York.

345. MAKSYMOWYCH, R. (1959). Quantitative analysis of leaf development in *Xanthium pensylvanicum. Am. J. Bot.*, **46**, 635–644.

346. MAKSYMOWYCH, R. (1963). Cell division and cell elongation in leaf development of *Xanthium pensylvanicum. Am. J. Bot.*, **50**, 891–901.

347. MAKSYMOWYCH, R., BLUM, M. K. and DEVLIN, R. G. (1966). Autoradiographic studies of the synthesis of nuclear DNA in various tissues during leaf development of *Xanthium pensylvanicum. Devl Biol.*, **13**, 250–265.

348. MAKSYMOWYCH, R. and ERICKSON, R. O. (1960). Development of the lamina in *Xanthium italicum* represented by the plastochron index. *Am. J. Bot.*, **47**, 451–459.

349. MAKSYMOWYCH, R. and WOCHOK, Z. S. (1969). Activity of marginal and plate meristems during leaf development of *Xanthium pensylvanicum. Am. J. Bot.*, **56**, 26–30.

350. MALLORY, T. E., CHIANG, S., CUTTER, E. G. and GIFFORD, E. M., Jr. (1970). Sequence and pattern of lateral root formation in five selected species. *Am. J. Bot.*, **57**, 800–809.

351. MARTIN, F. S. (1969). Pollen analysis and the scanning microscope. In *Scanning Electron Microscopy/1969. Proc. 2nd annual Scanning Electron Microscope Symp.*, 89–103. Chicago, Illinois.

352. MARTIN, F. W. (1969). Compounds from the stigmas of ten species. *Am. J. Bot.*, **56**, 1023–1027.

353. MARTIN, F. W. and ORTIZ, S. (1967). Anatomy of the stigma and style of sweet potato. *New Phytol.*, **66**, 109–113.

354. MCCALLUM, W. B. (1902). On the nature of the stimulus causing the change of form and structure in *Proserpinaca palustris. Bot. Gaz.*, **34**, 93–108.

355. MCCOMB, A. J. (1965). The control of elongation in *Callitriche* shoots by environment and gibberellic acid. *Ann. Bot., N.S.*, **29**, 445–458.

356. MCCULLY, M. E. and DALE, H. M. (1961). Heterophylly in *Hippuris*, a problem in identification. *Can. J. Bot.*, **39**, 1099–1116.

357. MELVILLE, R. and WRIGLEY, F. A. (1969). Fenestration in the leaves of *Monstera* and its bearing on the morphogenesis and colour patterns of leaves. *J. Linn. Soc. (Bot.)*, **62**, 1–16.

358. METCALFE, C. R. (1960). *Anatomy of the Monocotyledons. I. Gramineae.* Clarendon Press, Oxford.

359. METCALFE, C. R. (1963). Comparative anatomy as a modern botanical discipline. *Adv. bot. Res.*, **1**, 101–147.

360. METCALFE, C. R. and CHALK, L. (1950). *Anatomy of the Dicotyledons.* Vols. I and II. Clarendon Press, Oxford.

361. MIKSCHE, J. P. and GREENWOOD, M. (1966). Quiescent centre of the primary root of *Glycine max. New Phytol.*, **65**, 1–4.

362. MILLINGTON, W. F. (1963). Shoot tip abortion in *Ulmus americana. Am. J. Bot.*, **50**, 371–378.

363. MILTHORPE, F. L. and NEWTON, P. (1963). Studies on the expansion of the leaf surface. III. The influence of radiation on cell division and leaf expansion. *J. exp. Bot.*, **14**, 483–495.

364. MOHAN RAM, H. Y., RAM, M. and STEWARD, F. C. (1962). Growth and development of the banana plant. 3. A. The origin of the inflorescence and the development of the flowers. B. The structure and development of the fruit. *Ann. Bot., N.S.*, **26**, 657–673.

365. MOREY, P. R. and CRONSHAW, J. (1966). Induced structural changes in cambial derivatives of *Ulmus americana*. *Protoplasma*, **62**, 76–85.

366. MOREY, P. R. and CRONSHAW, J. (1968a). Developmental changes in the secondary xylem of *Acer rubrum* induced by various auxins and 2,3,5-tri-iodobenzoic acid. *Protoplasma*, **65**, 287–313.

367. MOREY, P. R. and CRONSHAW, J. (1968b). Induction of tension wood by 2,4-dinitrophenol and auxins. *Protoplasma*, **65**, 393–405.

367a. MOREY, P. R. and CRONSHAW, J. (1968c). Developmental changes in the secondary xylem of *Acer rubrum* induced by gibberellic acid, various auxins and 2,3,5-tri-iodobenzoic acid. *Protoplasma*, **65**, 315–326.

368. MORLAND, N. (1960). *Science in Crime Detection*. Emerson Books, Inc., New York.

369. MORRE, D. J. (1968). Cell wall dissolution and enzyme secretion during leaf abscission. *Pl. Physiol., Lancaster*, **43**, 1545–1559.

370. MOSELEY, M. F. (1961). Morphological studies of the Nymphaeaceae. II. The flower of *Nymphaea*. *Bot. Gaz.*, **122**, 233–259.

371. MOSS, G. I. and HESLOP-HARRISON, J. (1967). A cytochemical study of DNA, RNA and protein in the developing maize anther. II. Observations. *Ann. Bot., N.S.*, **31**, 555–572.

372. MURASHIGE, T. (1966). The deciduous behavior of a tropical plant, *Plumeria acuminata*. *Physiologia Pl.*, **19**, 348–355.

373. MURASHIGE, T., NAKANO, R. and TUCKER, D. P. H. (1967). Histogenesis and rate of nuclear change in *Citrus limon* tissue *in vitro*. *Phytomorphology*, **17**, 469–476.

374. NEČESANÝ, V. (1958). Effect of β-indoleacetic acid on the formation of reaction wood. *Phyton, B. Aires*, **11**, 117–127.

375. NEGBI, M. and LANG, A. (1961). Studies on orientation of cell division in *Hyoscyamus niger*. *Am. J. Bot.*, **48**, 529–530.

376. NELMES, B. J. and PRESTON, R. D. (1968). Wall development in apple fruits: a study of the life history of a parenchyma cell. *J. exp. Bot.*, **19**, 496–518.

377. NEWMAN, I. V. (1956). Pattern in meristems of vascular plants. I. Cell partition in living apices and in the cambial zone in relation to the concepts of initial cells and apical cells. *Phytomorphology*, **6**, 1–19.

378. NEWMAN, I. V. (1965). Pattern in the meristems of vascular plants. III. Pursuing the patterns in the apical meristem where no cell is a permanent cell. *J. Linn. Soc. (Bot.)*, **59**, 185–214.

379. NITSCH, C. (1968a). Induction *in vitro* de la floraison chez une plante de jours courts: *Plumbago indica* L. *Annls Sci. nat. (Bot.)*, Sér. **12, 9**, 1–92.

380. NITSCH, C. (1968b). Effects of growth substances on the induction of flowering of a short-day plant *in vitro*. In *Biochemistry and Physiology of Plant Growth Substances*, WIGHTMAN, F. and SETTERFIELD, G., 1385–1398. Runge Press Ltd., Ottawa.

381. NITSCH, J. P. (1950). Growth and morphogenesis of the strawberry as related to auxin. *Am. J. Bot.*, **37**, 211–215.

382. NITSCH, J. P. (1951). Growth and development *in vitro* of excised ovaries. *Am. J. Bot.*, **38**, 566–577.

382a. NITSCH, J. P. (1969). Experimental androgenesis in *Nicotiana*. *Phytomorphology*, **19**, 389–404.

383. NITSCH, J. P. and NITSCH, C. (1969). Haploid plants from pollen grains. *Science, N.Y.*, **163**, 85–87.

384. NOUGARÈDE, A. (1965). Organisation et fonctionnement du méristème apical des végétaux vasculaires. In *Travaux Dédiés à Lucien Plantefol*, 171–340. Masson et Cie., Paris.

385. NOUGARÈDE, A. (1967). Experimental cytology of the shoot apical cells during vegetative growth and flowering. *Internat. Rev. Cytol.*, **21**, 203–351.

386. NOUGARÈDE, A. and BERNIER, G. (1964). The intermediate phase in photoperiodic and cold-requiring plants; its signification. In *Differentiation of Apical Meristems and Some Problems of Plants*, 25–28. Academia, Praha.

387. NOUGARÈDE, A., BRONCHART, R., BERNIER, G. and RONDET, P. (1964). Comportement du méristème apical du *Perilla nankinensis* (Lour.) Decne. en relation avec les conditions photopériodiques. *Revue gén. Bot.*, **61**, 205–238.

388. NOUGARÈDE, A., GIFFORD, E. M., Jr. and RONDET, P. (1965). Cytohistological studies of the apical meristem of *Amaranthus retroflexus* under various photoperiodic regimes. *Bot. Gaz.*, **126**, 281–298.

389. OBATON, M. (1960). Les lianes ligneuses à structure anormale des forêts denses d'Afrique occidentale. *Annls Sci. nat. (Bot.)*, Sér. **12**, **1**, 1–220.

390. O'BRIEN, T. P., FEDER, N. and MCCULLY, M. E. (1965). Polychromatic staining of plant cell walls by Toluidine Blue O. *Protoplasma*, **59**, 367–373.

391. OHKUMA, K., LYON, J. L., ADDICOTT, F. T. and SMITH, O. E. (1963). Abscisin II, an abscission-accelerating substance from young cotton fruit. *Science, N.Y.*, **142**, 1592–1593.

392. O'NEILL, T. B. (1961). Primary vascular organization of *Lupinus* shoot. *Bot. Gaz.*, **123**, 1–9.

393. PANT, D. D. and KIDWAI, P. F. (1968). Structure and ontogeny of stomata in some Caryophyllaceae. *J. Linn. Soc. (Bot.)*, **60**, 309–314.

394. PATE, J. S. and GUNNING, B. E. S. (1969). Vascular transfer cells in angiosperm leaves. A taxonomic and morphological survey. *Protoplasma*, **68**, 135–156.

395. PAULET, P. (1965). Étude de la néoformation *in vitro* de bourgeons végétatifs et floreaux. *Revue gén. Bot.*, **72**, 697–792.

396. PECKET, R. C. (1957a). The initiation and development of lateral meristems in the pea root. I. The effect of young and of mature tissue. *J. exp. Bot.*, **8**, 172–180.

397. PECKET, R. C. (1957b). The initiation and development of lateral meristems in the pea root. II. The effect of indole-3-acetic acid. *J. exp. Bot.*, **8**, 181–194.

398. PELLEGRINI, O. (1957). Esperimenti chirurgici sul comportamento del meristema radicale di *Phaseolus vulgaris* L. *Delpinoa*, **10**, 187–199.

399. PETERSON, R. L. (1967). Differentiation and maturation of primary tissues in white mustard root tips. *Can. J. Bot.*, **45**, 319–331.

400. PHILIPSON, W. R. (1946). Studies in the development of the inflorescence. I. The capitulum of *Bellis perennis* L. *Ann. Bot.*, N.S., **10**, 257–270.

401. PHILIPSON, W. R. (1947a). Studies in the development of the inflorescence. II. The capitulum of *Succisa pratensis* Moench and *Dipsacus fullonum* L. *Ann. Bot., N.S.,* 11, 285–297.

402. PHILIPSON, W. R. (1947b). Studies in the development of the inflorescence. III. The thyrse of *Valeriana officinalis* L. *Ann. Bot., N.S.,* 11, 409–416.

403. PHILIPSON, W. R. (1948). Studies in the development of the inflorescence. IV. The capitula of *Hieracium boreale* Fries and *Dahlia gracilis* Ortg. *Ann. Bot., N.S.,* 12, 65–75.

404. PHILIPSON, W. R. (1949). The ontogeny of the shoot apex in dicotyledons. *Biol. Rev.,* 24, 21–50.

405. PHILIPSON, W. R. and BALFOUR, E. E. (1963). Vascular patterns in dicotyledons. *Bot. Rev.,* 29, 382–404.

406. PICKETT-HEAPS, J. D. and NORTHCOTE, D. H. (1966). Cell division in the formation of the stomatal complex of the young leaves of wheat. *J. Cell Sci.,* 1, 121–128.

407. POLLOCK, E. G. and JENSEN, W. A. (1967). Ontogeny and cytochemistry of the chalazal proliferating cells of *Capsella bursa-pastoris* (L.) Medic. *New Phytol.,* 66, 413–417.

408. POPHAM, R. A. (1951). Principal types of vegetative shoot apex organization in vascular plants. *Ohio J. Sci.,* 51, 249–270.

409. POPHAM, R. A. (1955). Levels of tissue differentiation in primary roots of *Pisum sativum. Am. J. Bot.,* 42, 529–540.

410. POPHAM, R. A. (1966). *Laboratory Manual for Plant Anatomy.* C. V. Mosby Company, Saint Louis.

411. PRAY, T. R. (1955a). Foliar venation of angiosperms. II. Histogenesis of the venation of *Liriodendron. Am. J. Bot.,* 42, 18–27.

412. PRAY, T. R. (1955b). Foliar venation of angiosperms. III. Pattern and histology of the venation of *Hosta. Am. J. Bot.,* 42, 611–618.

413. PRAY, T. R. (1955c). Foliar venation of angiosperms. IV. Histogenesis of the venation of *Hosta. Am. J. Bot.,* 42, 698–706.

414. PRAY, T. R. (1957). Marginal growth of leaves of monocotyledons: *Hosta, Maranta* and *Philodendron. Phytomorphology,* 7, 381–387.

415. PRIESTLEY, J. H. and SWINGLE, C. F. (1929). Vegetative propagation from the standpoint of plant anatomy. *U.S.D.A. Tech. Bull.,* 151.

416. PRITCHARD, H. N. (1964). A cytochemical study of embryo development in *Stellaria media. Am. J. Bot.,* 51, 472–479.

417. PURI, V. (1952). Placentation in angiosperms. *Bot. Rev.,* 18, 603–651.

418. RAGHAVAN, V. (1966). Nutrition, growth and morphogenesis of plant embryos. *Biol. Rev.,* 41, 1–58.

419. RAJU, M. V. S. (1969). Development of floral organs in the sites of leaf primordia in *Pinguicula vulgaris. Am. J. Bot.,* 56, 507–514.

420. RAJU, M. V. S., STEEVES, T. A. and NAYLOR, J. M. (1964). Developmental studies on *Euphorbia esula* L.: apices of long and short roots. *Can. J. Bot.,* 42, 1615–1628.

421. RAMJI, M. V. (1967). Morphology and ontogeny of the foliar venation of *Calophyllum inophyllum* L. *Aust. J. Bot.,* 15, 437–443.

422. REINHARD, E. (1956). Ein Vergleich zwischen diarchen und triarchen Wurzeln von *Sinapis alba. Z. Bot.,* 44, 505–514.

423. RICHARDS, F. J. (1948). The geometry of phyllotaxis and its origin. *Symp. Soc. exp. Biol.,* 2, 217–245.

424. RIJVEN, A. H. G. C. (1952). *In vitro* studies of the embryo of *Capsella bursa-pastoris. Acta bot. neerl.*, **1**, 157–200.

425. RIOPEL, J. L. (1964). Studies on the origin and potentiality of lateral root meristems. *A.S.B. Bull.*, **11**, 54.

426. RIOPEL, J. L. (1966). The distribution of lateral roots in *Musa acuminata* 'Gros Michel'. *Am. J. Bot.*, **53**, 403–407.

427. RIOPEL, J. L. (1969). Regulation of lateral root positions. *Bot. Gaz.*, **130**, 80–83.

428. RIOPEL, J. L. and STEEVES, T. A. (1964). Studies on the roots of *Musa acuminata* cv. Gros Michel. I. The anatomy and development of main roots. *Ann. Bot.*, N.S., **28**, 475–490.

429. ROBARDS, A. W. (1965). Tension wood and eccentric growth in crack willow (*Salix fragilis* L.). *Ann. Bot.*, N.S., **29**, 419–431.

430. ROBERTS, E. H. (1964a). The distribution of oxidation-reduction enzymes and effects of respiratory inhibitors and oxidising agents on dormancy in rice seed. *Physiologia Pl.*, **17**, 14–29.

431. ROBERTS, E. H. (1964b). A survey of the effects of chemical treatments on dormancy in rice seed. *Physiologia Pl.*, **17**, 30–43.

432. ROSADO-ALBERIO, J., WEIER, T. E. and STOCKING, C. R. (1968). Continuity of the chloroplast membrane systems in *Zea mays* L. *Pl. Physiol., Lancaster*, **43**, 1325–1331.

432a. ROSEN, W. G. and THOMAS, H. R. (1970). Secretory cells of lily pistils. I. *Am. J. Bot.*, **57**, 1108–1114.

433. ROSSO, S. W. (1968). The ultrastructure of chromoplast development in red tomatoes. *J. Ultrastruct. Res.*, **25**, 307–322.

434. SABLON, LECLERC DU (1885). Recherches sur la structure et la déhiscence des anthères. *Annls Sci. nat. (Bot.)*, Sér. **7**, **1**, 97–134.

435. SACHS, R. M. (1965). Stem elongation. *A. Rev. Pl. Physiol.*, **16**, 73–96.

436. SACHS, R. M. (1968). Control of intercalary growth in the scape of *Gerbera* by auxin and gibberellic acid. *Am. J. Bot.*, **55**, 62–68.

437. SACHS, R. M., BRETZ, C. F. and LANG, A. (1959). Shoot histogenesis: the early effects of gibberellin upon stem elongation in two rosette plants. *Am. J. Bot.*, **46**, 376–384.

438. SACHS, R. M. and LANG, A. (1959). Shoot histogenesis and the sub-apical meristem: the action of gibberellic acid, Amo-1618, and maleic hydrazide. In *Plant Growth Regulation*, 567–578. Iowa State Univ. Press.

439. SACHS, R. M., LANG, A., BRETZ, C. F. and ROACH, J. (1960). Shoot histogenesis: subapical meristematic activity in a caulescent plant and the action of gibberellic acid and Amo-1618. *Am. J. Bot.*, **47**, 260–266.

440. SACHS, T. (1968a). The role of the root in the induction of xylem differentiation in peas. *Ann. Bot.*, N.S., **32**, 391–399.

441. SACHS, T. (1968b). On the determination of the pattern of vascular tissue in peas. *Ann. Bot.*, N.S., **32**, 781–790.

442. SACHS, T. (1969). Polarity and the induction of organized vascular tissues. *Ann. Bot.*, N.S., **33**, 263–275.

443. SALISBURY, F. B. (1963). *The Flowering Process*. Pergamon Press, Oxford.

444. SALISBURY, F. B. (1969). *Xanthium strumarium* L. In *The Induction of Flowering*, EVANS, L. T., 14–61. Cornell Univ. Press, Ithaca, New York.

445. SANIO, C. (1864). Notiz über Verdickung des Hölzkorpers auf der Markseite bei *Tecoma radicans. Bot. Ztg*, **22**, 61.

446. SATINA, S. (1959). Chimeras. In *Blakeslee: the Genus Datura*, AVERY, A. G., SATINA, S. and RIETSEMA, J., 132–151. Ronald Press Co., New York.

447. SATINA, S., BLAKESLEE, A. F. and AVERY, A. G. (1940). Demonstration of the three germ layers in the shoot apex of *Datura* by means of induced polyploidy in periclinal chimeras. *Am. J. Bot.*, **27**, 895–905.

448. SAUTER, J. J. (1969). Autoradiographische Untersuchungen zur RNS-und Proteinsynthese in Pollenmutterzellen, jungen Pollen und Tapetumzellen während der Mikrosporogenese von *Paeonia tenuifolia* L. *Z. PflPhysiol.*, **61**, 1–19.

449. SAUTER, J. J. and MARQUARDT, H. (1967). Die Rolle des Nukleohistons bei der RNS- und Proteinsynthese während der Mikrosporogenese von *Paeonia tenuifolia* L. *Z. PflPhysiol.*, **58**, 126–137.

450. SCHENCK, F. (1893). *Beiträge zur Biologie und Anatomie der Lianen*. Vol. II. *Beiträge zur Anatomie der Lianen*. Gustav Fischer, Jena.

451. SCHMIDT, A. (1924). Histologische Studien an phanerogamen Vegetationspunkten. *Bot. Arch.*, **8**, 345–404.

452. SCHMIDT, B. L. and MILLINGTON, W. F. (1968). Regulation of leaf shape in *Proserpinaca palustris*. *Bull. Torrey bot. Club*, **95**, 264–286.

453. SCHNEIDER, H. (1968). The anatomy of *Citrus*. In *The Citrus Industry*, II. REUTHER, W., BATCHELOR, L. D., and WEBBER, H. J., 1–85. Univ. California, Div. Agri. Sci.

454. SCHÜEPP, O. (1917). Untersuchungen über Wachstum und Formwechsel von Vegetationspunkten. *Jb. wiss. Bot.*, **57**, 17–79.

455. SCHULZ, R. and JENSEN, W. A. (1968a). *Capsella* embryogenesis: the synergids before and after fertilization. *Am. J. Bot.*, **55**, 541–552.

456. SCHULZ, R. and JENSEN, W. A. (1968b). *Capsella* embryogenesis: the egg, zygote, and young embryo. *Am. J. Bot.*, **55**, 807–819.

457. SCHULZ, R. and JENSEN, W. A. (1968c). *Capsella* embryogenesis: the early embryo. *J. Ultrastruct. Res.*, **22**, 376–392.

458. SCHULZ, P. and JENSEN, W. A. (1969). *Capsella* embryogenesis: the suspensor and the basal cell. *Protoplasma*, **67**, 139–163.

459. SCHWABE, W. W. (1959). Some effects of environment and hormone treatment on reproductive morphogenesis in the Chrysanthemum. *J. Linn. Soc. (Bot.)*, **56**, 254–261.

460. SCOTT, P. C., MILLER, L. W., WEBSTER, B. D. and LEOPOLD, A. C. (1967). Structural changes during bean leaf abscission. *Am. J. Bot.*, **54**, 730–734.

461. SCURFIELD, G. (1964). The nature of reaction wood. IX. Anomalous cases of reaction anatomy. *Aust. J. Bot.*, **12**, 173–184.

462. SCURFIELD, G. and WARDROP, A. B. (1963). The nature of reaction wood. VII. Lignification in reaction wood. *Aust. J. Bot.*, **11**, 107–116.

463. SEN, S. (1968). An induced mutant of *Corchorus olitorius* L. with enhanced abscission rate. *Ann. Bot.*, *N.S.*, **32**, 863–866.

464. SEXTON, R. and SUTCLIFFE, J. F. (1969). The distribution of β-glycerophosphatase in young roots of *Pisum sativum* L. *Ann. Bot.*, *N.S.*, **33**, 407–419.

465. SHARMAN, B. C. and HITCH, P. A. (1967). Initiation of procambial strands in leaf primordia of bread wheat, *Triticum aestivum* L. *Ann. Bot.*, *N.S.*, **31**, 229–243.

466. SHUSHAN, S. and JOHNSON, M. A. (1955). The shoot apex and leaf of *Dianthus caryophyllus* L. *Bull. Torrey bot. Club*, **82**, 266–283.

467. SIMON, E. W. (1967). Types of leaf senescence. *Symp. Soc. exp. Biol.*, **21**, 215–230.
468. SIMS, W. L. and GLEDHILL, B. L. (1969). Ethrel effects on sex expression, and growth development in pickling cucumbers. *Calif. Agric.*, **23**, 2, 4–6.
469. SINNOTT, E. W. (1936). A developmental analysis of inherited shape differences in cucurbit fruits. *Am. Nat.*, **70**, 245–254.
470. SINNOTT, E. W. (1939). A developmental analysis of the relation between cell size and fruit size in cucurbits. *Am. J. Bot.*, **26**, 179–189.
471. SINNOTT, E. W. (1944). Cell polarity and the development of form in cucurbit fruits. *Am. J. Bot.*, **31**, 388–391.
472. SINNOTT, E. W. (1945). The relation of growth to size in cucurbit fruits. *Am. J. Bot.*, **32**, 439–446.
473. SINNOTT, E. W. (1963). *The Problem of Organic Form.* Yale University Press, New Haven.
474. SLACK, C. R. and HATCH, M. D. (1967). Comparative studies on the activity of carboxylases and other enzymes in relation to the new pathway of photosynthetic carbon dioxide fixation in tropical grasses. *Biochem. J.*, **103**, 660–665.
475. SLADE, B. F. (1957). Leaf development in relation to venation as shown in *Cercis siliquastrum* L., *Prunus serrulata* Lindl. and *Acer pseudoplatanus* L. *New Phytol.*, **56**, 281–300.
476. SMITH, D. L. (1966). Development of the inflorescence in *Carex. Ann. Bot.*, N.S., **30**, 475–486.
477. SMITH, D. L. (1967). The experimental control of inflorescence development in *Carex. Ann. Bot.*, N.S., **31**, 19–30.
478. SMITH, D. L. (1969). The role of leaves and roots in the control of inflorescence development in *Carex. Ann. Bot.*, N.S., **33**, 505–514.
479. SMITH, O.E. (1969). Changes in abscission-accelerating substances with development of cotton fruit. *New Phytol.*, **68**, 313–322.
479a. SNOW, M. (1951). Experiments on spirodistichous shoot apices. I. *Phil. Trans. R. Soc. B*, **235**, 131–162.
480. SNOW, M. and SNOW, R. (1933). Experiments on phyllotaxis. II. The effect of displacing a primordium. *Phil. Trans. R. Soc. B*, **222**, 353–400.
481. SNOW, M. and SNOW, R. (1942). The determination of axillary buds. *New Phytol.*, **41**, 13–22.
482. SNOW, M. and SNOW, R. (1947). On the determination of leaves. *New Phytol.*, **46**, 5–19.
483. SNOW, R. (1935). Activation of cambial growth by pure hormones. *New Phytol.*, **34**, 347–360.
483a. SNOW, R. (1954). Phyllotaxis of flowering teasels. *New Phytol.*, **53**, 99–107.
483b. SNOW, R. (1955). Problems of phyllotaxis and leaf determination. *Endeavour*, **14**, 190–199.
484. SÖDING, H. (1926). Über den Einfluss der junge Infloreszenz auf das Wachstum ihres Schaftes. *Jb. wiss. Bot.*, **65**, 611–635.
485. SOETIARTO, S. R. and BALL, E. (1969a). Ontogenetical and experimental studies of the floral apex of *Portulaca grandiflora*. 1. Histology of transformation of the shoot apex into the floral apex. *Can. J. Bot.*, **47**, 133–140.

486. SOETIARTO, S. R. and BALL, E. (1969b). Ontogenetical and experimental studies of the floral apex of *Portulaca grandiflora*. 2. Bisection of the meristem in successive stages. *Can. J. Bot.*, **47**, 1067-1076.

487. SOLEREDER, H. and MEYER, F. J. (1928). *Systematische Anatomie der Monokotyledonen*. Gebrüder Borntraeger, Berlin.

488. SOROKIN, H. P., MATHUR, S. N. and THIMANN, K. V. (1962). The effects of auxins and kinetin on xylem differentiation in the pea epicotyl. *Am. J. Bot.*, **49**, 444-454.

489. SOUÈGES, R. (1919). Les premières divisions de l'oeuf et les differenciations du suspenseur chez le *Capsella bursa-pastoris* Moench. *Annls Sci. nat.*, *Bot.*, Sér. **10**, **1**, 1-28.

490. SOUÈGES, R. (1938-51). *Embryogénie et Classification*. Hermann & Cie., Paris.

491. SPICHINGER, J. U. (1969). Isolation und Charakterisierung von Sphärosomen und Glyoxisomen aus Tabakendosperm. *Planta*, **89**, 56-75.

492. SRIVASTAVA, L. M. (1966). On the fine structure of the cambium of *Fraxinus americana* L. *J. Cell Biol.*, **31**, 79-93.

493. STEEVES, T. A., HICKS, M. A., NAYLOR, J. M. and RENNIE, P. (1969). Analytical studies on the shoot apex of *Helianthus annuus*. *Can. J. Bot.*, **47**, 1367-1375.

494. STERLING, C. (1945). Growth and vascular development in the shoot apex of *Sequoia sempervirens* (Lamb.) Endl. II. Vascular development in relation to phyllotaxis. *Am. J. Bot.*, **32**, 380-386.

495. STEVENSON, G. (1970). *The Biology of Fungi, Bacteria and Viruses*. 2nd edition. Edward Arnold, London.

496. STREET, H. E. and MCGREGOR, S. M. (1952). The carbohydrate nutrition of tomato roots. III. The effects of external sucrose concentration on the growth and anatomy of excised roots. *Ann. Bot.*, *N.S.*, **16**, 185-205.

497. STREET, H. E. and ÖPIK, H. (1970). *The Physiology of Flowering Plants: Their Growth and Development*. Edward Arnold, London.

498. SUNDERLAND, N. (1960). Cell division and expansion in the growth of the leaf. *J. exp. Bot.*, **11**, 68-80.

499. SUSSEX, I. M. and CLUTTER, M. E. (1960). A study of the effect of externally supplied sucrose on the morphology of excised fern leaves *in vitro*. *Phytomorphology*, **10**, 87-99.

500. TEPFER, S. S. (1953). Floral anatomy and ontogeny in *Aquilegia formosa* var. *truncata* and *Ranunculus repens*. *Univ. Calif. Publs Bot.*, **25**, 513-648.

501. TEPFER, S. S. and CHESSIN, M. (1959). Effects of tobacco mosaic virus on early leaf development in tobacco. *Am. J. Bot.*, **46**, 496-509.

502. TEPFER, S. S., GREYSON, R. I., CRAIG, W. R., and HINDMAN, G. L. (1963). *In vitro* culture of floral buds of *Aquilegia*. *Am. J. Bot.*, **50**, 1035-1045.

503. TEPFER, S. S., KARPOFF, A. J. and GREYSON, R. I. (1966). Effects of growth substances on excised floral buds of *Aquilegia*. *Am. J. Bot.*, **53**, 148-157.

504. TETLEY, M. (1930). A study of the anatomical development of the apple and some observations on the 'pectic constituents' of the cell walls. *J. Pomol.*, **8**, 153-172.

505. THIMANN, K. V. (1957). Discussion in: Decennial review conference on tissue culture. *J. natn. Cancer Inst.*, **19**, 660.

506. THODAY, D. (1939). The interpretation of plant structure. *Adv. Sci.*, **1**, 84–104.

507. THOMAS, R. G. (1963). Floral induction and the stimulation of cell division in *Xanthium*. *Science, N.Y.*, **140**, 54–56.

508. THOMPSON, J. and CLOWES, F. A. L. (1968). The quiescent centre and rates of mitosis in the root meristem of *Allium sativum*. *Ann. Bot.*, *N.S.*, **32**, 1–13.

509. THOMSON, B. F. and MILLER, P. M. (1962). The role of light in histogenesis and differentiation in the shoot of *Pisum sativum*. I. The apical region. *Am. J. Bot.*, **49**, 303–310.

510. TOMLINSON, P. B. (1961). *Anatomy of the Monocotyledons*. II. *Palmae*. Clarendon Press, Oxford.

511. TOMLINSON, P. B. (1964). Stem structure in arborescent monocotyledons. In *The Formation of Wood in Forest Trees*, ZIMMERMANN, M. H., 65–86. Academic Press, New York.

512. TOMLINSON, P. B. and MOORE, H. E., Jr. (1968). Inflorescence in *Nannorrhops ritchiana* (Palmae). *J. Arnold Arbor.*, **49**, 16–34.

513. TOMLINSON, P. B. and ZIMMERMANN, M. H. (1966a). Vascular bundles in palm stems—their bibliographic evolution. *Proc. Am. phil. Soc.*, **110**, 174–181.

514. TOMLINSON, P. B. and ZIMMERMANN, M. H. (1966b). Anatomy of the palm *Rhapis excelsa*. III. Juvenile phase. *J. Arnold Arbor.*, **47**, 301–312.

515. TOMLINSON, P. B. and ZIMMERMANN, M. H. (1967). The 'wood' of monocotyledons. *Bull. int. Ass. Wood Anat.*, 1967/2, 4–24.

516. TOMLINSON, P. B. and ZIMMERMANN, M. H. (1969). Vascular anatomy of monocotyledons with secondary growth—an introduction. *J. Arnold Arbor.*, **50**, 159–179.

517. TORREY, J. G. (1955). On the determination of vascular patterns during tissue differentiation in excised pea roots. *Am. J. Bot.*, **42**, 183–198.

518. TORREY, J. G. (1957). Auxin control of vascular pattern formation in regenerating pea root meristems grown *in vitro*. *Am. J. Bot.*, **44**, 859–870.

519. TORREY, J. G. (1961). The initiation of lateral roots. In *Recent Advances in Botany*, 808–812. Univ. Toronto Press.

520. TORREY, J. G. (1963). Cellular patterns in developing roots. *Symp. Soc. exp. Biol.*, **17**, 285–314.

521. TORREY, J. G. (1965). Physiological bases of organization and development in the root. *Handb. PflPhysiol.*, **15**, 1, 1256–1327.

522. TORREY, J. G. and LOOMIS, R. S. (1967a). Auxin-cytokinin control of secondary vascular tissue formation in isolated roots of *Raphanus*. *Am. J. Bot.*, **54**, 1098–1106.

522a. TORREY, J. G. and LOOMIS, R. S. (1967b). Ontogenetic studies of vascular cambium formation in excised roots of *Raphanus sativus* L. *Phytomorphology*, **17**, 401–409.

523. TROMP, J. (1968). Flower-bud formation and shoot growth in apple as affected by shoot orientation. *Acta bot. neerl.*, **17**, 212–220.

524. TUCKER, S. C. (1959). Ontogeny of the inflorescence and the flower in *Drimys winteri* var. *chilensis*. *Univ. Calif. Publs Bot.*, **30**, 257–336.

525. TUCKER, S. C. (1966). The gynoecial vascular supply in *Caltha*. *Phytomorphology*, **16**, 339–342.

526. TUCKER, S. C. (1968). Meristem, determinate. In *McGraw-Hill Yearbook of Science and Technology* (1968), 250–253. McGraw-Hill, New York.

527. TUCKER, S. C. and GIFFORD, E. M., Jr. (1964). Carpel vascularization of *Drimys lanceolata*. *Phytomorphology*, **14**, 197–203.
528. TUCKER, S. C. and GIFFORD, E. M., Jr. (1966a). Organogenesis in the carpellate flower of *Drimys lanceolata*. *Am. J. Bot.*, **53**, 433–442.
529. TUCKER, S. C. and GIFFORD, E. M., Jr. (1966b). Carpel development in *Drimys lanceolata*. *Am. J. Bot.*, **53**, 671–678.
530. TUKEY, H. B. and YOUNG, J. O. (1939). Histological study of the developing fruit of the sour cherry. *Bot. Gaz.*, **100**, 723–749.
531. TUKEY, H. B. and YOUNG, J. O. (1942). Gross morphology and histology of developing fruit of the apple. *Bot. Gaz.*, **104**, 3–25.
532. TURING, A. M. (1952). The chemical basis of morphogenesis. *Phil. Trans. R. Soc. B*, **237**, 37–72.
533. VAN FLEET, D. S. (1950). A comparison of histochemical and anatomical characteristics of the hypodermis with the endodermis in vascular plants. *Am. J. Bot.*, **37**, 721–725.
534. VAN FLEET, D. S. (1961). Histochemistry and function of the endodermis. *Bot. Rev.*, **27**, 165–220.
534a. VAN ITERSON, G. (1907). Mathematische und mikroskopische-anatomische Studien über Blattstellungen. Fischer, Jena.
535. VAN OVERBEEK, J. and CRUZADO, H. J. (1948). Flower formation in the pineapple plant by geotropic stimulation. *Am. J. Bot.*, **35**, 410–412.
536. VARNER, J. E. and RAM CHANDRA, G. (1964). Hormonal control of enzyme synthesis in barley endosperm. *Proc. natn. Acad. Sci. U.S.A.*, **52**, 100–106.
537. VASIL, I. K. (1967). Physiology and cytology of anther development. *Biol. Rev.*, **42**, 327–373.
538. VAUGHAN, J. G. (1956). The seed coat structure of *Brassica integrifolia* (West) O. E. Schulz var. *carinata* (A. Br.). *Phytomorphology*, **6**, 363–367.
539. VAUGHAN, J. G. (1959). The testa of some *Brassica* seeds of oriental origin. *Phytomorphology*, **9**, 107–110.
540. VAUGHAN, J. G. (1968). Seed anatomy and taxonomy. *Proc. Linn. Soc. Lond.*, **179**, 251–255.
541. WAISEL, Y. and FAHN, A. (1965). The effects of environment on wood formation and cambial activity in *Robinia pseudacacia* L. *New Phytol.*, **64**, 436–442.
542. WAISEL, Y., LIPHSCHITZ, N. and ARZEE, T. (1967). Phellogen activity in *Robinia pseudacacia* L. *New Phytol.*, **66**, 331–335.
543. WAISEL, Y., NOAH, I. and FAHN, A. (1966). Cambial activity in *Eucalyptus camaldulensis* Dehn. II. The production of phloem and xylem elements. *New Phytol.*, **65**, 319–324.
544. WALLENSTEIN, A. and ALBERT, L. S. (1963). Plant morphology: its control in *Proserpinaca* by photoperiod, temperature and gibberellic acid. *Science, N.Y.*, **140**, 998–1000.
545. WARDLAW, C. W. (1928). Size in relation to internal morphology. 3. The vascular system of roots. *Trans. R. Soc. Edinb.*, **56**, 19–55.
546. WARDLAW, C. W. (1943). Experimental and analytical studies of pteridophytes. I. Preliminary observations on the development of buds on the rhizome of the ostrich fern (*Matteuccia struthiopteris* Tod.). *Ann. Bot., N.S.*, **7**, 171–184.
547. WARDLAW, C. W. (1944). Experimental and analytical studies of pteridophytes. III. Stelar morphology: the initial differentiation of vascular tissue. *Ann. Bot., N.S.*, **8**, 173–188.

548. WARDLAW, C. W. (1947). Experimental investigations of the shoot apex of *Dryopteris aristata* Druce. *Phil. Trans. R. Soc. B*, **232**, 343–384.

549. WARDLAW, C. W. (1949). Experimental and analytical studies of pteridophytes. XIV. Leaf formation and phyllotaxis in *Dryopteris aristata* Druce. *Ann. Bot., N.S.*, **13**, 163–198.

550. WARDLAW, C. W. (1950). The comparative investigation of apices of vascular plants by experimental methods. *Phil. Trans. R. Soc. B.*, **234**, 583–604.

551. WARDLAW, C. W. (1952). The effect of isolating the apical meristem in *Echinopsis, Nuphar, Gunnera* and *Phaseolus*. *Phytomorphology*, **2**, 240–242.

552. WARDLAW, C. W. (1953). Comparative observations on the shoot apices of vascular plants. *New Phytol.*, **52**, 195–209.

553. WARDLAW, C. W. (1955). *Embryogenesis in Plants*. Methuen, London.

554. WARDLAW, C. W. (1956). Experimental and analytical studies of pteridophytes. XXXII. Further investigations on the effect of undercutting fern leaf primordia. *Ann. Bot., N.S.*, **20**, 121–132.

555. WARDLAW, C. W. (1957a). The reactivity of the apical meristem as ascertained by cytological and other techniques. *New Phytol.*, **56**, 221–229.

556. WARDLAW, C. W. (1957b). Experimental and analytical studies of pteridophytes. XXXVII. A note on the inception of microphylls. *Ann. Bot., N.S.*, **21**, 427–437.

557. WARDLAW, C. W. (1961). Growth and development of the inflorescence and flower. In *Growth in Living Systems*, ZARROW, M. X., 491-523. Basic Books, Inc., New York.

558. WARDLAW, C. W. (1965a). *Organization and Evolution in Plants*. Longmans, London.

559. WARDLAW, C. W. (1965b). Physiology of embryonic development in cormophytes. *Handb. PflPhysiol.*, **15**, 1, 844–965.

560. WARDLAW, C. W. (1965c). The organization of the shoot apex. *Handb. PflPhysiol.*, **15**, 1, 966–1076.

561. WARDLAW, C. W. (1968). *Morphogenesis in Plants*. Methuen, London.

562. WARDROP, A. B. (1964). The reaction anatomy of arborescent angiosperms. In *The Formation of Wood in Forest Trees*, ZIMMERMANN, M. H., 405–456. Academic Press, New York.

563. WARDROP, A. B. (1965). The formation and function of reaction wood. In *Cellular Ultrastructure of Woody Plants*, CÔTÉ, W. A., Jr., 371–390. Syracuse University Press.

564. WARDROP, A. B. and DAVIES, G. W. (1964). The nature of reaction wood. VIII. The structure and differentiation of compression wood. *Aust. J. Bot.*, **12**, 24–38.

565. WAREING, P. F. (1956). Photoperiodism in woody plants. *A. Rev. Pl. Physiol.*, **7**, 191–214.

566. WAREING, P. F. (1958). Interaction between indoleacetic acid and gibberellic acid in cambial activity. *Nature, Lond.*, **181**, 1744–1745.

567. WAREING, P. F. (1969). Germination and dormancy. In *The Physiology of Plant Growth and Development*, WILKINS, M. B., 605–644. McGraw-Hill, London.

568. WAREING, P. F., HANNEY, C. E. A. and DIGBY, J. (1964). The role of endogenous hormones in cambial activity and xylem differentiation. In

The Formation of Wood in Forest Trees, ZIMMERMANN, M. H., 323–344. Academic Press, New York.

569. WAREING, P. F. and NASR, T. (1958). Gravimorphism in trees. Effects of gravity on growth, apical dominance and flowering in fruit trees. *Nature, Lond.*, **182**, 379–380.

570. WAREING, P. F. and ROBERTS, D. L. (1956). Photoperiodic control of cambial activity in *Robinia pseudacacia* L. *New Phytol.*, **55**, 356–366.

571. WAREING, P. F. and SETH, A. K. (1967). Ageing and senescence in the whole plant. *Symp. Soc. exp. Biol.*, **21**, 543–558.

572. WARREN WILSON, J. and WARREN WILSON, P. M. (1961). The position of regenerating cambia—a new hypothesis. *New Phytol.*, **60**, 63–73.

573. WARREN WILSON, P. M. and WARREN WILSON, J. (1961). Cambium formation in wounded Solanaceous stems. *Ann. Bot., N.S.*, **25**, 104–115.

574. WARREN WILSON, P. M. and WARREN WILSON, J. (1963). Cambial regeneration in approach grafts between petioles and stems. *Aust. J. biol. Sci.*, **16**, 6–18.

575. WEBSTER, B. D. (1968). Anatomical aspects of abscission. *Pl. Physiol., Lancaster*, **43**, 1512–1544.

576. WEBSTER, B. D. (1969). Abscission. In *McGraw-Hill Yearbook of Science and Technology*, 1969, 85–88. McGraw-Hill, New York.

577. WEISS, C. and VAADIA, Y. (1965). Kinetin-like activity in root apices of sunflower plants. *Life Sci.*, **4**, 1323–1326.

578. WETMORE, R. H. and GARRISON, R. (1966). The morphological ontogeny of the leafy shoot. In *Trends in Plant Morphogenesis*, CUTTER, E. G., et al., 187–199. Longmans, London.

579. WETMORE, R. H., GIFFORD, E. M., Jr. and GREEN, M. C. (1959). Development of vegetative and floral buds. In *Photoperiodism and Related Phenomena in Plants and Animals*, WITHROW, R. B., 255–273.

580. WETMORE, R. H. and RIER, J. P. (1963). Experimental induction of vascular tissues in callus of angiosperms. *Am. J. Bot.*, **50**, 418–430.

581. WHALEY, W. G. and WHALEY, C. Y. (1942). A developmental analysis of inherited leaf patterns in *Tropaeolum*. *Am. J. Bot.*, **29**, 195–200.

582. WHITMORE, T. C. (1962). Studies in systematic bark morphology. II. General features of bark construction in Dipterocarpaceae. *New Phytol.*, **61**, 208–220.

583. WHITMORE, T. C. (1963). Studies in systematic bark morphology. IV. The bark of beech, oak and sweet chestnut. *New Phytol.*, **62**, 161–169.

584. WREN, M. J. and HANNAY, J. W. (1963). Ageing in roots of groundsel (*Senecio vulgaris*, L.). I. The root system of seedlings cultured aseptically in darkness. *New Phytol.*, **62**, 249–256.

585. WYLIE, R. B. (1951). Principles of foliar organization shown by sun–shade leaves from ten species of deciduous dicotyledonous trees. *Am. J. Bot.*, **38**, 355–361.

586. YOUNG, B. S. (1954). The effects of leaf primordia on differentiation in the stem. *New Phytol.*, **53**, 445–460.

587. ZEEVAART, J. A. D. (1969). Perilla. In *The Induction of Flowering*, EVANS, L. T., 116–155. Cornell Univ. Press, Ithaca, New York.

588. ZIMMERMANN, M. H. and TOMLINSON, P. B. (1969). The vascular system in the axis of *Dracaena fragrans* (Agavaceae). I. Distribution and development of primary strands. *J. Arnold Arbor.*, **50**, 370–383.

589. ZIMMERMANN, M. H., WARDROP, A. B. and TOMLINSON, P. B. (1968). Tension wood in aerial roots of *Ficus benjamina* L. *Wood Sci. Technol.*, **2**, 95–104.

Index

Major entries (page numbers) are shown in **bold** type, those referring to illustrations in *italic* type